A PHOTOGRAPHIC ATLAS FOR

Anatomy & Physiology

A PHOTOGRAPHIC ATLAS FOR

Anatomy & Physiology

Nora Hebert
Red Rocks Community College

Ruth E. Heisler
University of Colorado at Boulder

with

Jett Chinn
Cañada College and College of Marin

Karen M. Krabbenhoft
University of Wisconsin – Madison, School of Medicine & Public Health

Olga Malakhova
University of Florida College of Medicine, Gainesville

PEARSON

Boston Columbus Indianapolis New York San Francisco Hoboken
Amsterdam Cape Town Dubai London Madrid Milan Munich Paris Montréal Toronto
Delhi Mexico City São Paulo Sydney Hong Kong Seoul Singapore Taipei Tokyo

Editor-in-Chief: Serina Beauparlant
Project Editor: Nicole Taché
Editorial Assistant: Arielle Grant
Text Permissions Project Manager: Rachel Youdelman
Text Permissions Specialist: Timothy Nicholls
Managing Editor: Mike Early
Production Project Manager: Caroline Ayres
Production Management: Mary Tindle, S4Carlisle Publishing Services
Copyeditor: Mike Rossa

Compositor: S4Carlisle Publishing Services
Design Manager: Derek Bacchus
Interior and Cover Designer: Tammy Newnam
Illustrators: Precision Graphics, Kristina Seymour; Pearson Digital Pub, Corin Skidds
Photo Permissions Management: Donna Kalal
Photo Researcher: Bill Smith Group
Senior Manufacturing Buyer: Stacey Weinberger
Senior Marketing Manager: Allison Rona
Senior Anatomy & Physiology Specialist: Derek Perrigo

Cover Photo Credits: Karen Krabbenhoft, Pearson Science; Model courtesy of 3B Scientific®
Interior Photo Credits: page C-1

Library of Congress Cataloging-in-Publication Data
Hebert, Nora.
A photographic atlas for anatomy & physiology / Nora Hebert [and four others]. — First edition.
pages cm
ISBN 978-0-321-86925-8 — ISBN 0-321-86925-7 1. Anatomy—Atlases. 2. Physiology—Atlases. I. Title. II. Title: Photographic atlas for anatomy and physiology.
QM25.H43 2015
612.0022'2—dc23

2014006369

12 18

PEARSON

www.pearsonhighered.com

ISBN 10: 0-321-86925-7;
ISBN 13: 978-0-321-86925-8
(Student edition standalone)

ISBN 10: 0-321-96141-2;
ISBN 13: 978-0-321-96141-9
(Instructor Review Copy)

PREFACE

A Photographic Atlas for Anatomy & Physiology is a new visual study tool that will help students learn and identify key anatomical structures. Featuring photos from *Practice Anatomy Lab*™ 3.0 and other sources, the Atlas includes over 300 images of the most commonly used specimens in the undergraduate Anatomy & Physiology or Human Anatomy lab course: cadavers, light micrographs, cats, and anatomical models from leading manufacturers such as 3B Scientific®, SOMSO®, and Denoyer-Geppert Science Company. Photos have been enlarged for in-depth viewing, and numbered structure keys allow students to quickly test their knowledge of anatomy. Spreads with corresponding cadaver and model images also help students to better learn and identify structures.

The Atlas is composed of 13 chapters, organized by body system, including a final chapter with cat dissection photos. In each chapter, students will first explore gross anatomy, as seen on cadavers and anatomical models, and then conclude with relevant histological images.

The Atlas is available for students as an unbound, three-hole punched product (ISBN: 0-321-86925-7), and can also be placed in a value pack (ISBN: 0-321-96142-0). The Atlas can also be customized by chapter and by figure through Pearson's Custom Group.

Acknowledgments

A special thank you to the following individuals who contributed photographs to *Practice Anatomy Lab*™ 3.0:

Samuel Chen, *Moraine Valley Community College*

Larry DeLay, *Tri-Power Performance, Inc.*

Stephen W. Downing, *University of Minnesota, Medical School*

Lisa M. J. Lee, *The Ohio State University, College of Medicine*

Winston Charles Poulton, *University of Florida College of Medicine, Gainesville*

Leif Saul, *University of Colorado at Boulder*

Renn Sminkey, *Creative Digital Visions, LLC*

Michael J. Timmons, *Moraine Valley Community College*

Eksel Perez and Peter Westra, *San Francisco, CA*

Nina Zanetti, *Siena College*

We would also like to thank the following instructors for their expertise and thoughtful feedback in reviewing the Atlas:

Joslyn Ahlgren, *University of Florida*

Robert Michael Anson, *Community College of Baltimore County*

Ellen E. Beidler, *Shelton State Community College*

Tamyra Carmona, *Cosumnes River College*

Samuel Chen, *Moraine Valley Community College*

James A. Collier, *Truckee Meadows Community College*

David Conley, *Washington State University*

Judith D'Aleo, *Plymouth State University*

Audra Day, *South Plains College*

Lynn Gargan, *Tarrant County College*

Kimberly Kerr, *Troy University*

Christine Kornet, *University of Hawaii at Hilo*

John H. Wilkins, *Ball State University*

Matthew J. Wood, *Lake-Sumter State College*

CONTENTS

CHAPTER 3 The Skeletal System **21**

Axial Skeleton

Appendicular Skeleton

Peripheral Nervous System

Autonomic Nervous System

Special Senses

Tissues of the Nervous System

CHAPTER 6 The Endocrine System 125

Gross Anatomy of the Endocrine System

Tissues of the Endocrine System

CHAPTER 7 The Cardiovascular System 133

Heart

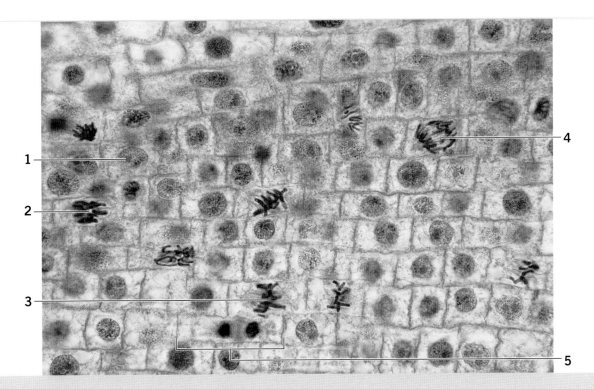

FIGURE 1.1 Mitosis, Onion Root Tip, 1000×

1. Interphase
2. Prophase
3. Metaphase
4. Anaphase
5. Telophase

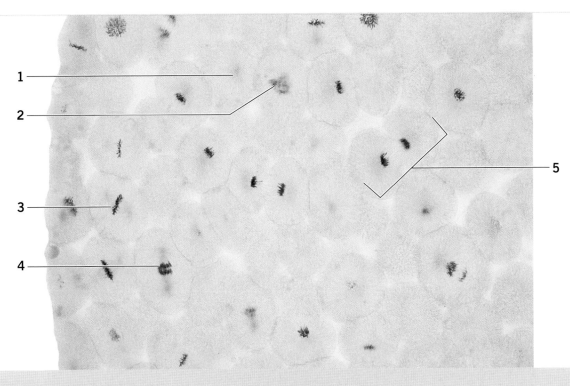

FIGURE 1.2 Mitosis, Whitefish Blastula, 400×

1. Interphase
2. Prophase
3. Metaphase
4. Anaphase
5. Telophase

Epithelial Tissue

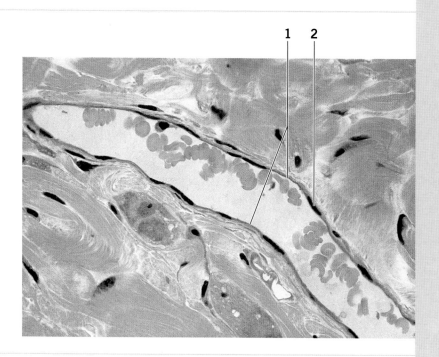

FIGURE 1.3 Simple Squamous Epithelium, Venule, 1000×

1. Simple squamous epithelium
2. Nucleus

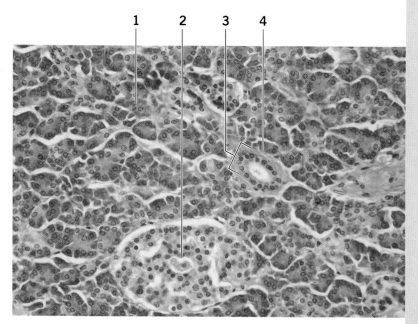

FIGURE 1.4 Simple Cuboidal Epithelium, Pancreatic Duct, 400×

1. Acinar tissue
2. Pancreatic islet
3. Duct lined with simple cuboidal epithelium
4. Basement membrane

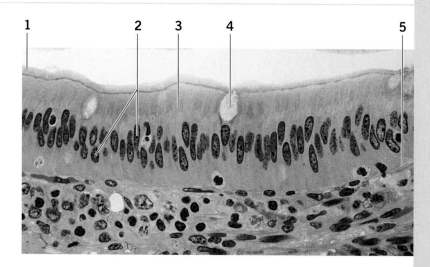

FIGURE 1.5 Simple Columnar Epithelium, Small Intestine, 1000×

1. Microvilli
2. Nuclei
3. Simple columnar epithelium
4. Goblet cell
5. Basement membrane

Epithelial Tissue

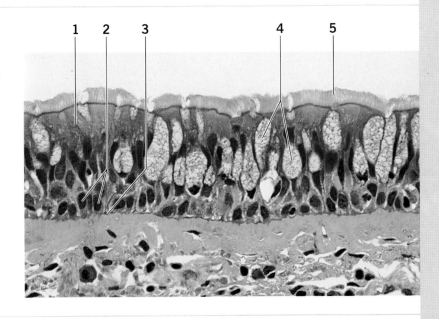

FIGURE 1.6 Pseudostratified Columnar Epithelium, Trachea, 1000×

1. Pseudostratified columnar epithelium
2. Nuclei
3. Basement membrane
4. Goblet cells
5. Cilia

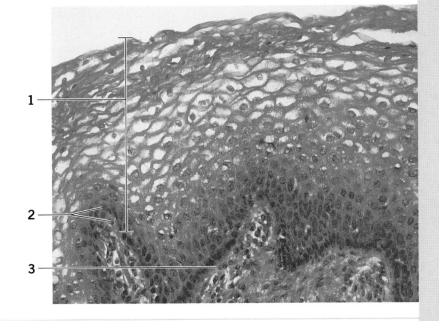

FIGURE 1.7 Stratified Squamous Epithelium, Nonkeratinized, Vagina, 400×

1. Nonkeratinized stratified squamous epithelium
2. Nuclei
3. Basement membrane

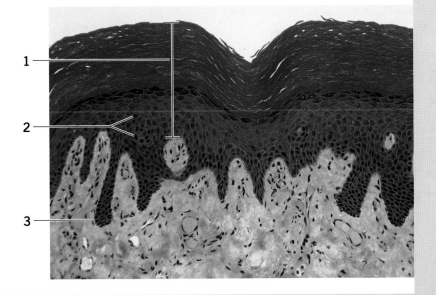

FIGURE 1.8 Stratified Squamous Epithelium, Keratinized, Thick Skin, 400×

1. Stratified squamous epithelium, keratinized
2. Nuclei
3. Basement membrane

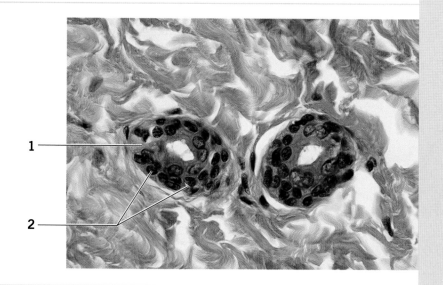

FIGURE 1.9 Stratified Cuboidal Epithelium, Sweat Gland in Dermis, 1000×

1. Stratified cuboidal epithelium
2. Nuclei

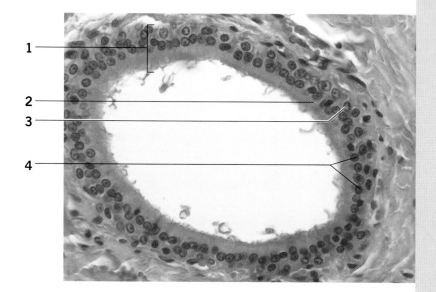

FIGURE 1.10 Stratified Columnar Epithelium, Duct in Parotid Gland, 1000×

1. Stratified columnar epithelium
2. Columnar epithelial cell
3. Cuboidal basal cell
4. Nuclei

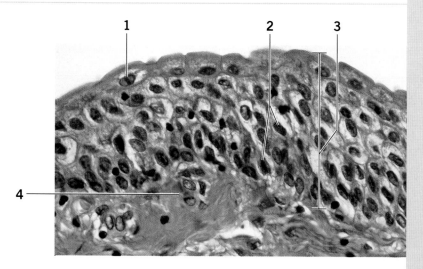

FIGURE 1.11 Transitional Epithelium, Urinary Bladder, 1000×

1. Umbrella cell
2. Nuclei
3. Transitional epithelium
4. Basement membrane

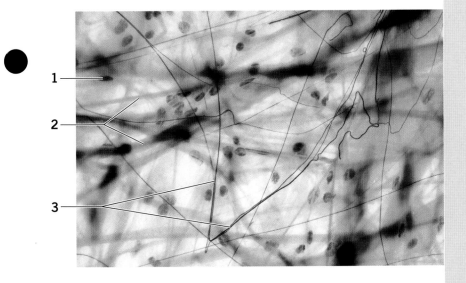

FIGURE 1.12 Areolar Connective Tissue, Mesentery, 400×

1. Fibroblast nucleus
2. Collagen fibers
3. Elastic fibers

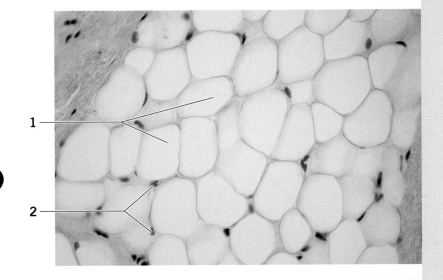

FIGURE 1.13 Adipose Tissue, Hypodermis, 400×

1. Adipocytes (fat cells)
2. Nuclei

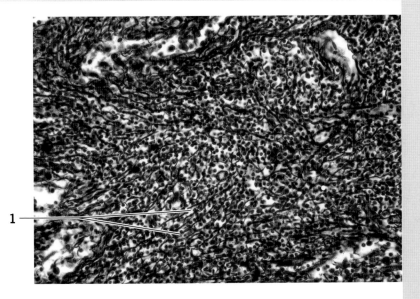

FIGURE 1.14 Reticular Connective Tissue, Lymph Node, 1000×

1. Reticular fibers

FIGURE 1.15 Dense Regular Connective Tissue, Tendon, 100×

1. Dense regular connective tissue (tendon)
2. Skeletal muscle

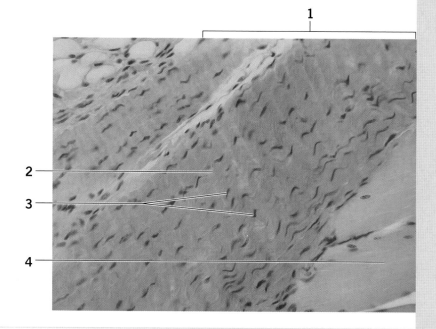

FIGURE 1.16 Dense Regular Connective Tissue, Tendon, 400×

1. Dense regular connective tissue (tendon)
2. Collagen fibers
3. Fibroblast nuclei
4. Skeletal muscle

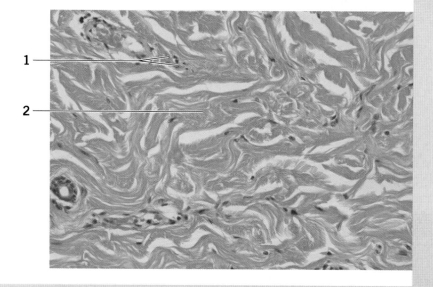

FIGURE 1.17 Dense Irregular Connective Tissue, Dermis, Thick Skin, 400×

1. Fibroblast nuclei
2. Collagen fibers

Connective Tissue

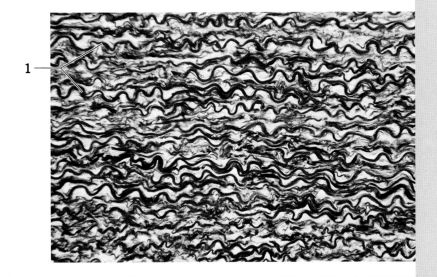

FIGURE 1.18 Elastic Connective Tissue, Aorta, 400×

1. Elastic sheets

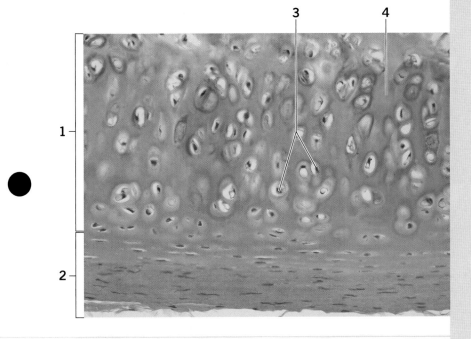

FIGURE 1.19 Hyaline Cartilage, Trachea, 400×

1. Hyaline cartilage
2. Perichondrium
3. Chondrocytes in lacunae
4. Matrix

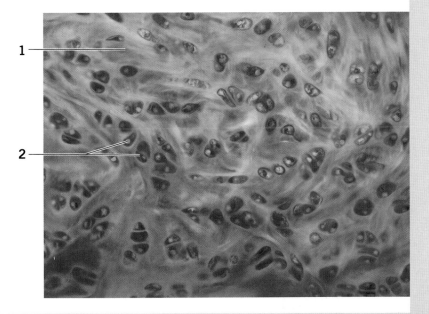

FIGURE 1.20 Fibrocartilage, Pubic Symphysis, 400×

1. Matrix
2. Chondrocytes in lacunae

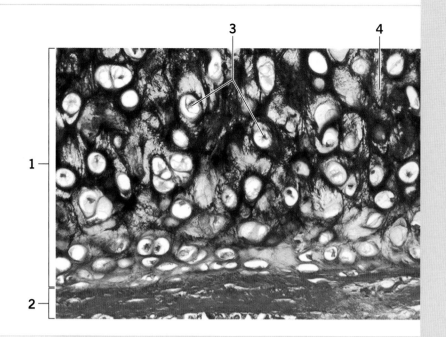

FIGURE 1.21 Elastic Cartilage, Epiglottis, 400×

1. Elastic cartilage
2. Perichondrium
3. Chondrocytes in lacunae
4. Matrix

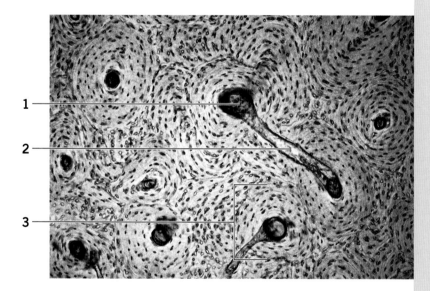

FIGURE 1.22 Compact Bone, Ground, 100×

1. Central canal
2. Perforating canal
3. Osteon (Haversian system)

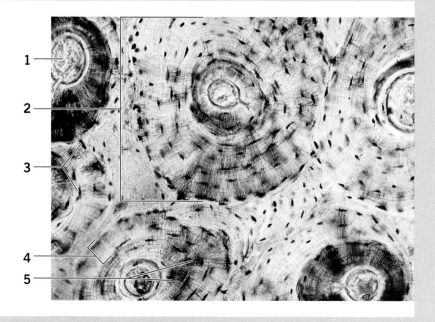

FIGURE 1.23 Compact Bone, Ground, 200×

1. Central canal
2. Osteon
3. Lacunae
4. Lamella
5. Canaliculi

Connective Tissue

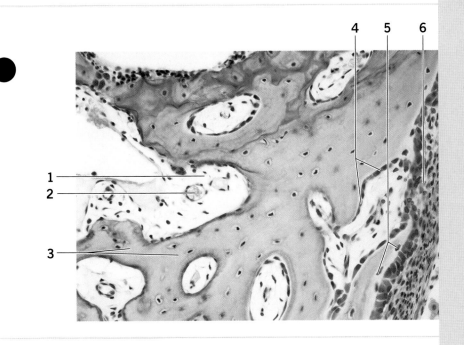

FIGURE 1.24 Developing Bone, Fetal Humerus, 400×

1. Bone marrow
2. Blood vessel
3. Trabeculae of spongy bone
4. Osteoblasts
5. Osteocytes
6. Periosteum

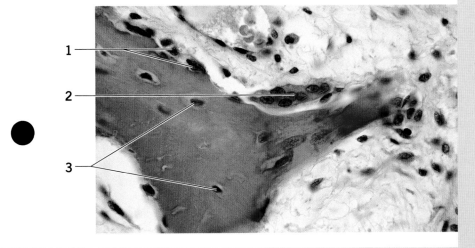

FIGURE 1.25 Spongy Bone, 1000×

1. Osteoblasts
2. Osteoclast
3. Osteocytes in lacunae

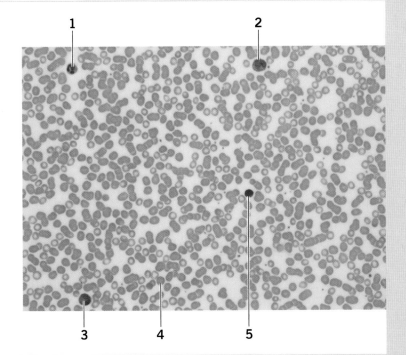

FIGURE 1.26 Blood Smear, 400×

1. Neutrophil
2. Monocyte
3. Eosinophil
4. Red blood cell
5. Lymphocyte

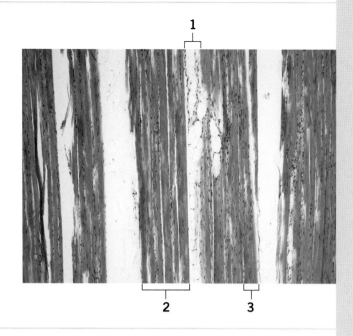

FIGURE 1.27 Skeletal Muscle Tissue, Longitudinal Section, 100×

1. Perimysium
2. Fascicle
3. Skeletal muscle fiber

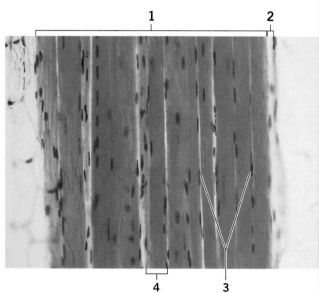

FIGURE 1.28 Skeletal Muscle Tissue, Longitudinal Section, 400×

1. Fascicle
2. Perimysium
3. Nuclei
4. Skeletal muscle fiber

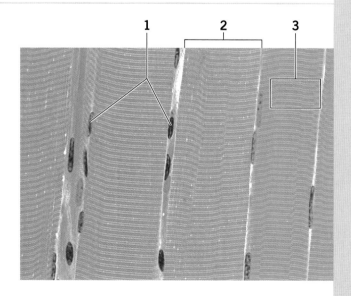

FIGURE 1.29 Skeletal Muscle Tissue, Longitudinal Section, 1200×

1. Nuclei
2. Skeletal muscle fiber
3. Striations

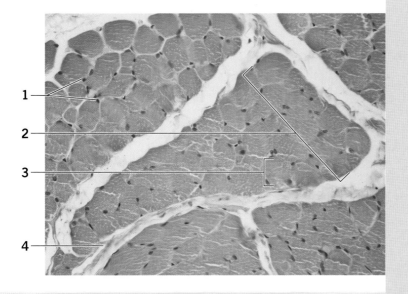

FIGURE 1.30 Skeletal Muscle Tissue, Cross Section, 400×

1. Nuclei
2. Fascicle
3. Skeletal muscle fiber
4. Perimysium

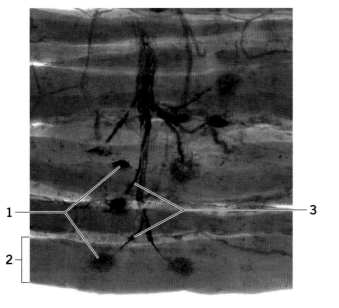

FIGURE 1.31 Neuromuscular Junction, 400×

1. Motor end plates
2. Skeletal muscle fiber
3. Motor axon terminals

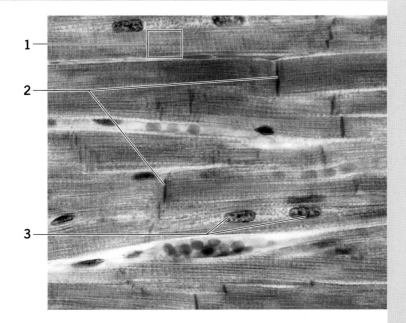

FIGURE 1.32 Cardiac Muscle, Longitudinal Section, 1000×

1. Striations
2. Intercalated discs
3. Nuclei

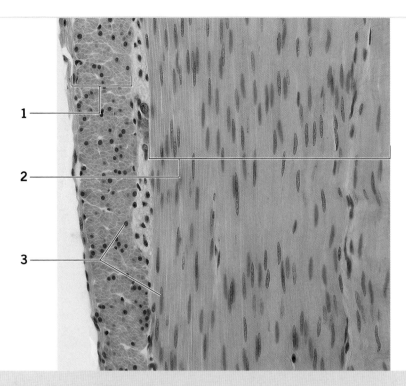

FIGURE 1.33 Smooth Muscle Tissue, Duodenum, Cross Section, 400×

1. Longitudinal layer of smooth muscle
2. Circular layer of smooth muscle
3. Nuclei

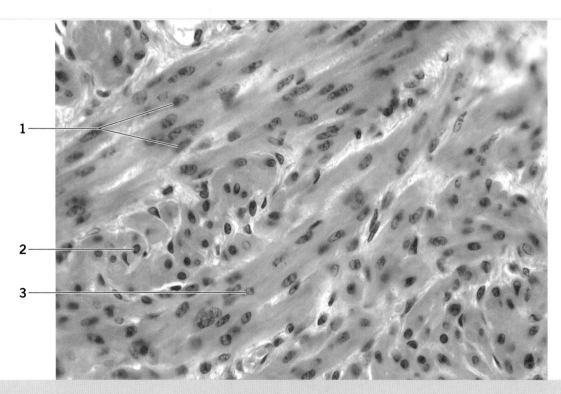

FIGURE 1.34 Smooth Muscle Tissue, Uterus, Cross Section, 1000×

1. Nuclei
2. Circular fiber of smooth muscle
3. Longitudinal fiber of smooth muscle

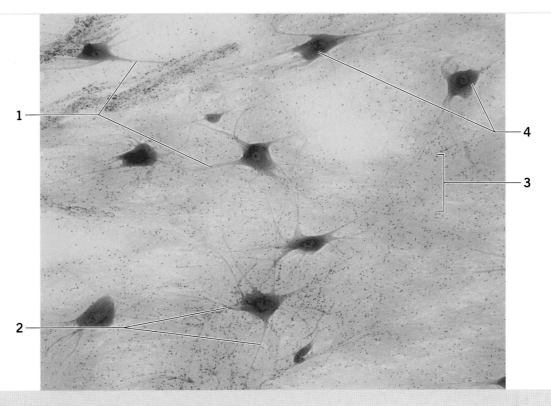

FIGURE 1.35 Multipolar Neuron Smear, Ventral Horn of Spinal Cord, 100×

1. Axons   **2.** Dendrites   **3.** Neuroglia   **4.** Cell bodies

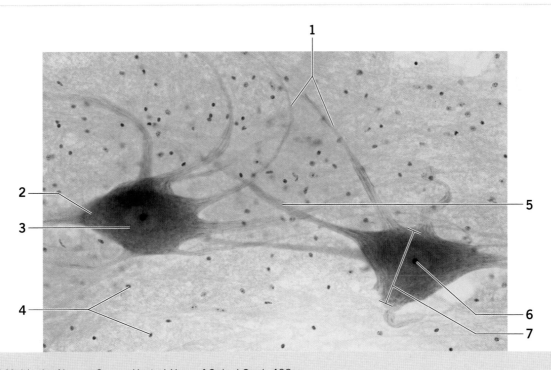

FIGURE 1.36 Multipolar Neuron Smear, Ventral Horn of Spinal Cord, 400×

1. Dendrites   **3.** Nucleus   **5.** Axon   **7.** Cell body
2. Axon hillock   **4.** Neuroglia   **6.** Nucleolus

FIGURE 1.37 Nerve Fibers, Teased Preparation, 400×

1. Myelinated nerve fiber
2. Node of Ranvier (myelin sheath gap)

FIGURE 1.38 Peripheral Nerve, Longitudinal Section, 40×

1. Fascicles
2. Perineurium

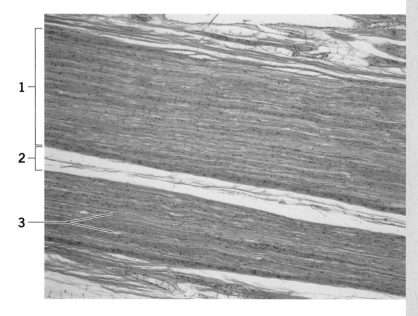

FIGURE 1.39 Peripheral Nerve, Longitudinal Section, 100×

1. Fascicle
2. Perineurium
3. Axons

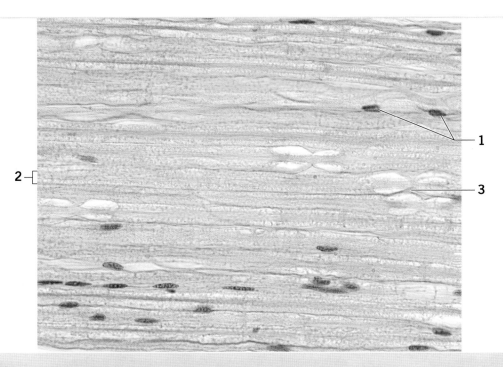

FIGURE 1.40 Peripheral Nerve, Longitudinal Section, 600×

1. Schwann cell nuclei

2. Myelinated nerve fiber

3. Node of Ranvier (myelin sheath gap)

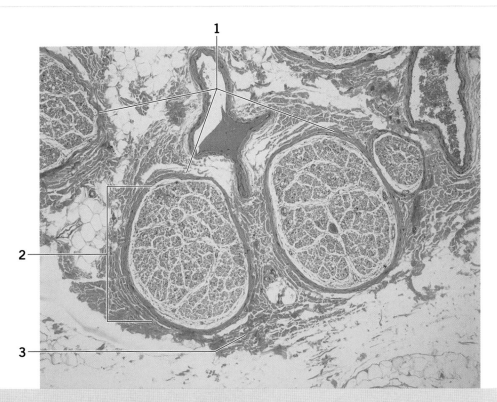

FIGURE 1.41 Peripheral Nerve, Cross Section, 100×

1. Perineurium

2. Fascicle

3. Epineurium

FIGURE 2.1 Layers of Skin

1. Hypodermis   **2.** Dermis   **3.** Epidermis

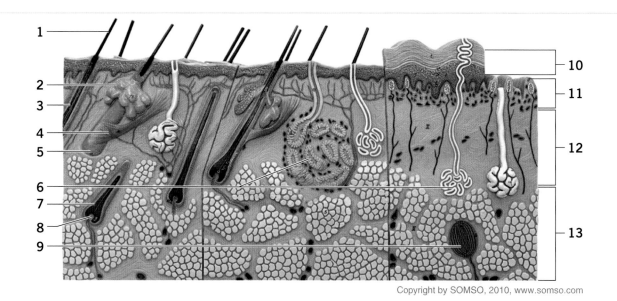

Copyright by SOMSO, 2010, www.somso.com

FIGURE 2.2 Skin Model

1. Hair shaft
2. Sebaceous gland
3. Hair root
4. Arrector pili muscle
5. Hair follicle

6. Sudoriferous (sweat) glands
7. Hair bulb
8. Hair papilla

9. Pacinian (lamellar) corpuscle
10. Epidermis
11. Papillary layer of dermis

12. Reticular layer of dermis
13. Hypodermis

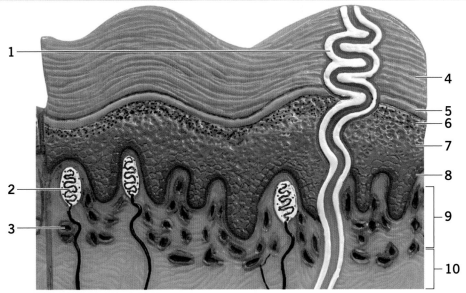

FIGURE 2.3 Epidermis of Thick Skin

1. Duct of sudoriferous gland
2. Meissner's (tactile) corpuscle
3. Blood vessel
4. Stratum corneum
5. Stratum lucidum
6. Stratum granulosum
7. Stratum spinosum
8. Stratum basale
9. Papillary layer of dermis
10. Reticular layer of dermis

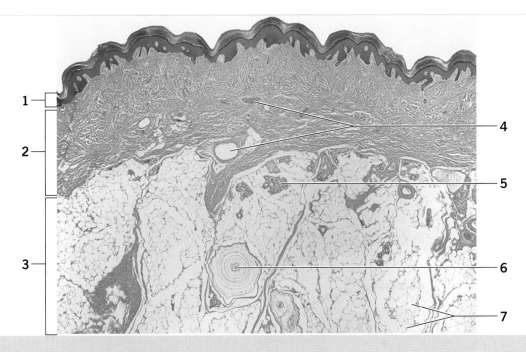

FIGURE 2.4 Thick Skin, Longitudinal Section, 40×

1. Epidermis
2. Dermis
3. Hypodermis
4. Blood vessels
5. Eccrine sweat gland
6. Pacinian (lamellar) corpuscle
7. Adipocytes

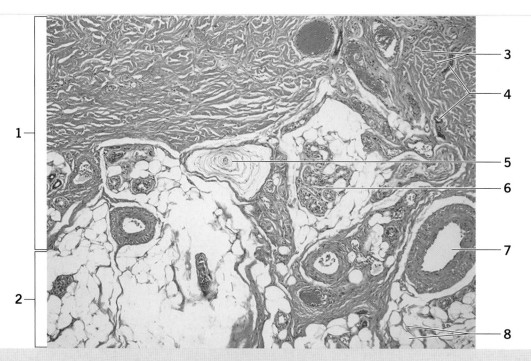

FIGURE 2.5 Dermis and Hypodermis of Thick Skin, Longitudinal Section, 100×

1. Dermis
2. Hypodermis
3. Collagen fibers
4. Eccrine ducts
5. Pacinian (lamellar) corpuscle
6. Eccrine sweat gland
7. Blood vessel
8. Adipocytes

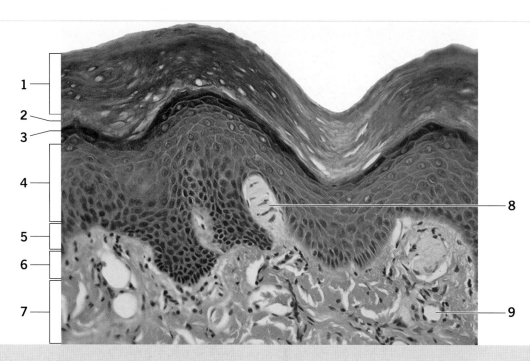

FIGURE 2.6 Epidermis and Dermis of Thick Skin, Longitudinal Section, 400×

1. Stratum corneum
2. Stratum lucidum
3. Stratum granulosum
4. Stratum spinosum
5. Stratum basale
6. Papillary layer of dermis
7. Reticular layer of dermis
8. Meissner's (tactile) corpuscle
9. Blood vessel

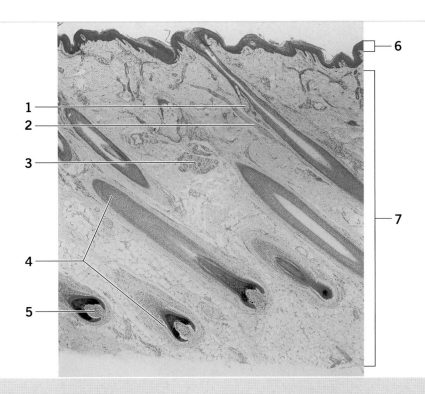

FIGURE 2.7 Hair Follicles in Thin Skin, Longitudinal Section, 40×

1. Sebaceous gland
2. Arrector pili muscle
3. Eccrine sweat gland
4. Hair follicles
5. Papilla
6. Epidermis
7. Dermis and hypodermis

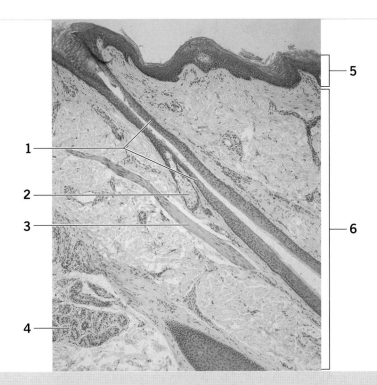

FIGURE 2.8 Hair Follicle in Thin Skin, Longitudinal Section, 100×

1. Follicle wall
2. Sebaceous gland
3. Arrector pili muscle
4. Eccrine sweat gland
5. Epidermis
6. Dermis

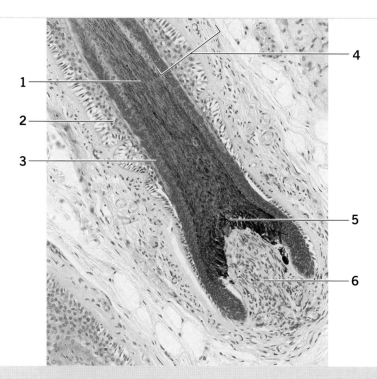

FIGURE 2.9 Hair Follicle in Thin Skin, Longitudinal Section, 150×

1. Hair root
2. External root sheath
3. Internal root sheath
4. Follicle wall
5. Matrix
6. Papilla

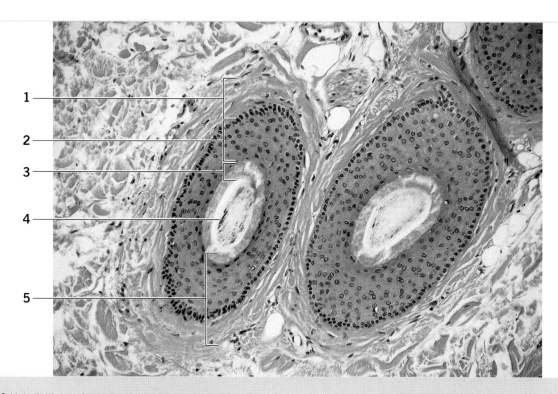

FIGURE 2.10 Hair Follicles, Cross Section, 200×

1. Connective tissue root sheath
2. External root sheath
3. Internal root sheath
4. Hair shaft
5. Follicle wall

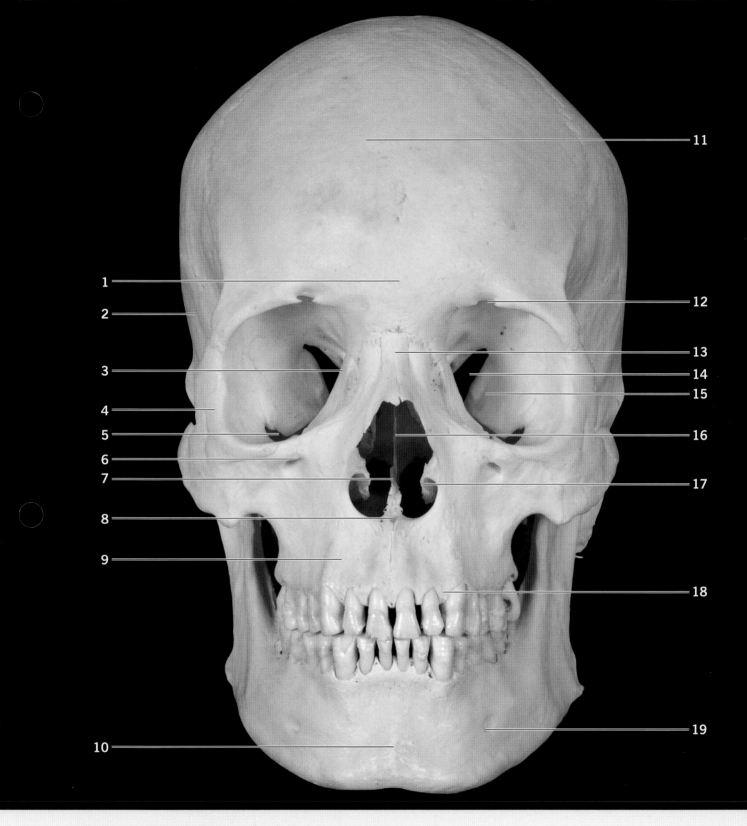

FIGURE 3.1 Skull, Anterior View

1. Glabella
2. Temporal bone
3. Lacrimal bone
4. Zygomatic bone
5. Inferior orbital fissure
6. Infraorbital foramen
7. Vomer bone
8. Anterior nasal spine
9. Maxilla
10. Mandible
11. Frontal bone
12. Supraorbital notch
13. Nasal bone
14. Superior orbital fissure
15. Sphenoid bone
16. Perpendicular plate of ethmoid bone
17. Inferior nasal concha
18. Alveolar margin
19. Mental foramen

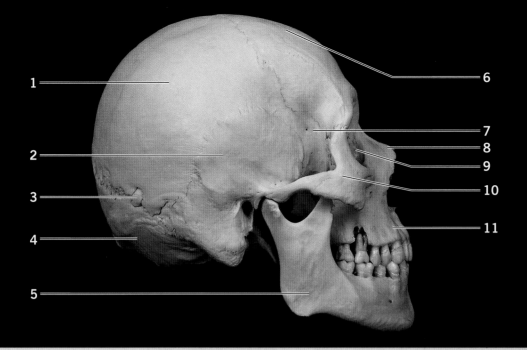

FIGURE 3.2 Skull, Lateral View

1. Parietal bone
2. Temporal bone
3. Sutural bone
4. Occipital bone
5. Mandible
6. Frontal bone
7. Sphenoid bone
8. Nasal bone
9. Lacrimal bone
10. Zygomatic bone
11. Maxilla

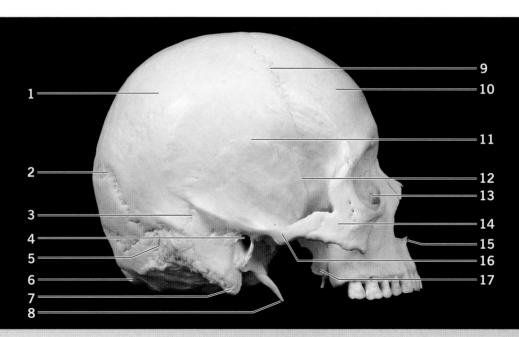

FIGURE 3.3 Cranium, Lateral View

1. Parietal bone
2. Lambdoid suture
3. Parietomastoid suture
4. External acoustic meatus
5. Occipitomastoid suture
6. External occipital protuberance
7. Mastoid process
8. Styloid process
9. Coronal suture
10. Frontal bone
11. Squamous suture
12. Sphenosquamous suture
13. Lacrimal fossa
14. Temporal process
15. Anterior nasal spine
16. Zygomatic process
17. Pterygoid process

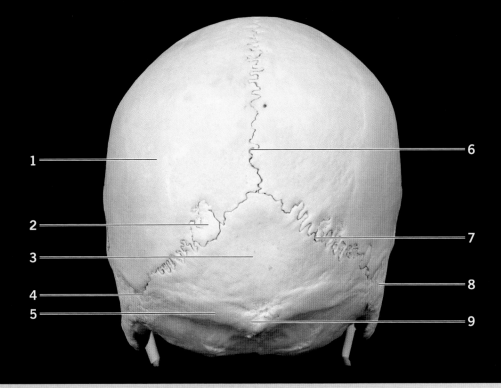

FIGURE 3.4 Cranium, Posterior View

1. Parietal bone	**4.** Occipitomastoid suture	**7.** Lambdoid suture	**9.** External occipital protuberance
2. Sutural bone	**5.** Superior nuchal line	**8.** Temporal bone	
3. Occipital bone	**6.** Sagittal suture		

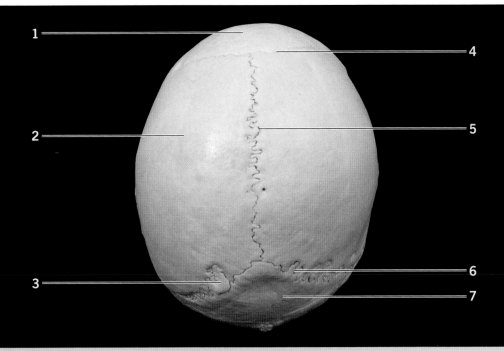

FIGURE 3.5 Skull, Superior View

1. Frontal bone	**3.** Sutural bone	**5.** Sagittal suture	**7.** Occipital bone
2. Parietal bone	**4.** Coronal suture	**6.** Lambdoid suture	

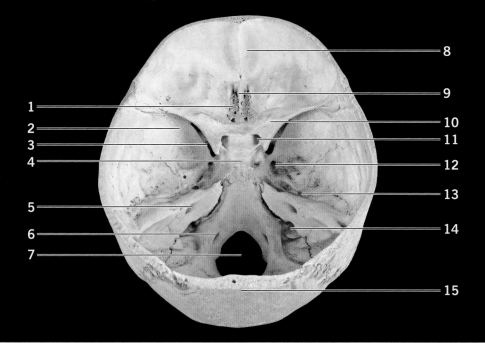

FIGURE 3.6 Cranial Cavity, Posterosuperior View

1. Cribriform plate
2. Greater wing of sphenoid
3. Superior orbital fissure
4. Sella turcica
5. Internal acoustic meatus
6. Hypoglossal canal
7. Foramen magnum
8. Frontal bone
9. Crista galli
10. Lesser wing of sphenoid
11. Optic canal
12. Foramen rotundum
13. Temporal bone
14. Jugular foramen
15. Occiptal bone

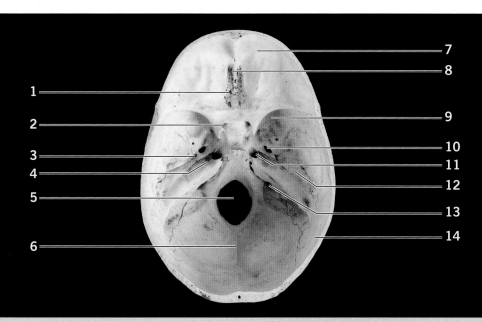

FIGURE 3.7 Cranial Cavity, Superior View

1. Olfactory foramen of cribriform plate
2. Optic canal
3. Foramen spinosum
4. Carotid canal
5. Foramen magnum
6. Occipital bone
7. Frontal bone
8. Crista galli
9. Sphenoid bone
10. Foramen ovale
11. Foramen lacerum
12. Temporal bone
13. Jugular foramen
14. Parietal bone

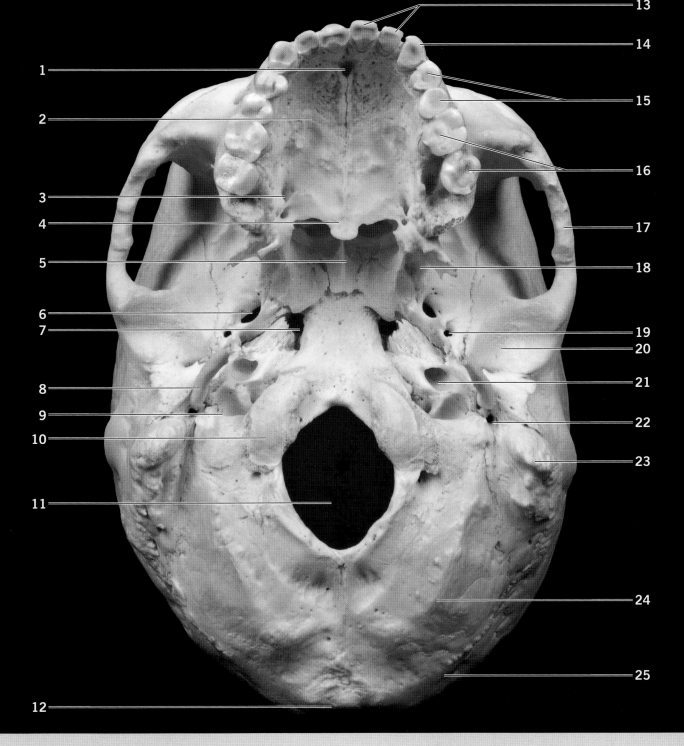

FIGURE 3.8 Cranium, Inferior View

1. Incisive fossa
2. Maxilla
3. Greater palatine foramen
4. Palatine bone
5. Vomer
6. Foramen ovale

7. Foramen lacerum
8. Styloid process
9. Jugular foramen
10. Occipital condyle
11. Foramen magnum
12. External occipital protuberance

13. Incisors
14. Canine
15. Premolars
16. Molars
17. Zygomatic arch
18. Pterygoid process
19. Foramen spinosum

20. Mandibular fossa
21. Carotid canal
22. Stylomastoid foramen
23. Mastoid process
24. Inferior nuchal line
25. Superior nuchal line

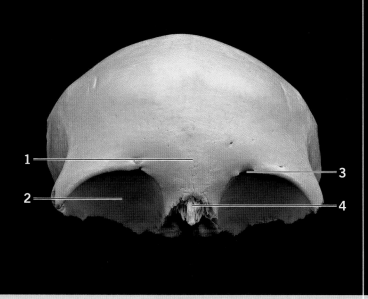

FIGURE 3.9 Frontal Bone, Anterior View

1. Glabella
2. Orbital surface
3. Supraorbital notch
4. Nasal spine

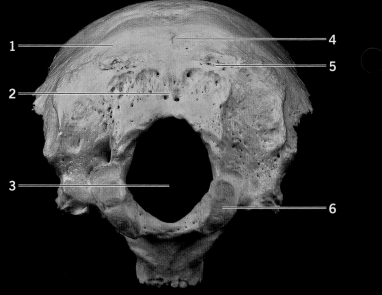

FIGURE 3.10 Occipital Bone, Inferior View

1. Superior nuchal line
2. External occipital crest
3. Foramen magnum
4. External occipital protuberance
5. Inferior nuchal line
6. Occipital condyle

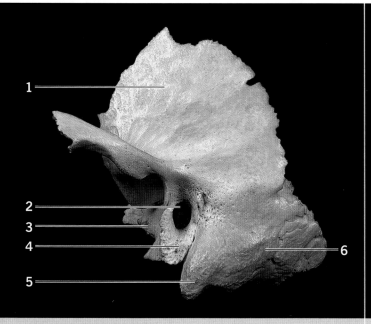

FIGURE 3.11 Left Temporal Bone, Lateral View

1. Squamous region
2. External acoustic meatus
3. Petrous region
4. Tympanic region
5. Mastoid process
6. Mastoid region

FIGURE 3.12 Left Temporal Bone, Medial View

1. Squamous region
2. Petrous region
3. Mastoid region
4. Internal acoustic meatus
5. Mastoid process

Axial Skeleton

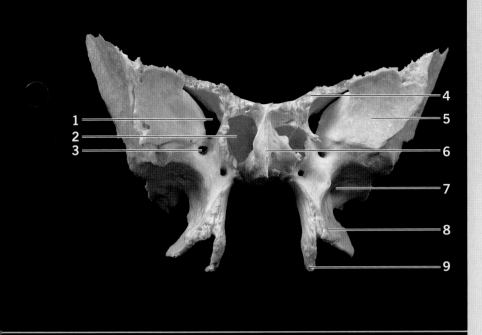

FIGURE 3.13 Sphenoid Bone, Anterior View

1. Superior orbital fissure
2. Sphenoidal sinus
3. Foramen rotundum
4. Lesser wing
5. Greater wing
6. Body
7. Foramen ovale
8. Lateral pterygoid plate
9. Medial pterygoid plate

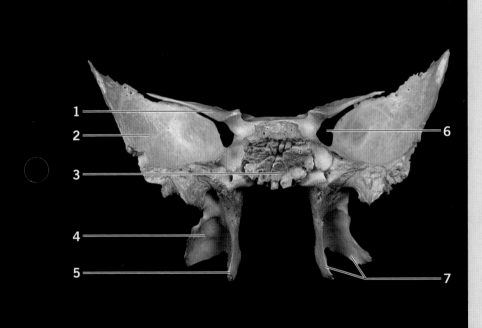

FIGURE 3.14 Sphenoid Bone, Posterior View

1. Lesser wing
2. Greater wing
3. Body
4. Lateral pterygoid plate
5. Medial pterygoid plate
6. Superior orbital fissure
7. Pterygoid process

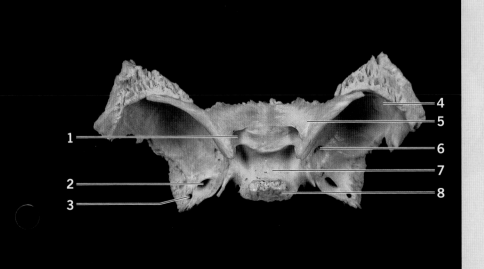

FIGURE 3.15 Sphenoid Bone, Superior View

1. Optic canal
2. Foramen ovale
3. Foramen spinosum
4. Greater wing
5. Lesser wing
6. Foramen rotundum
7. Sella turcica
8. Body

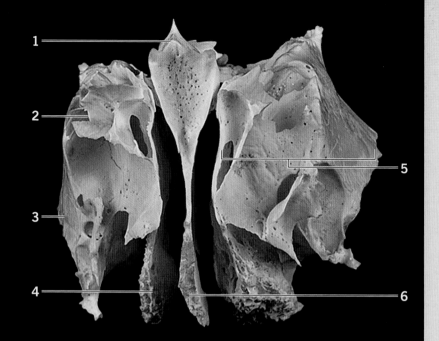

FIGURE 3.16 Ethmoid, Anterior View

1. Crista galli
2. Ethmoidal air cell
3. Orbital plate
4. Middle nasal concha
5. Lateral mass
6. Perpendicular plate

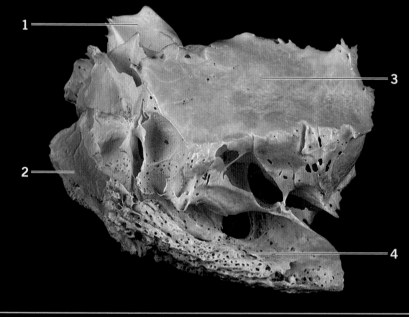

FIGURE 3.17 Ethmoid, Lateral View

1. Crista galli
2. Perpendicular plate
3. Orbital plate
4. Middle nasal concha

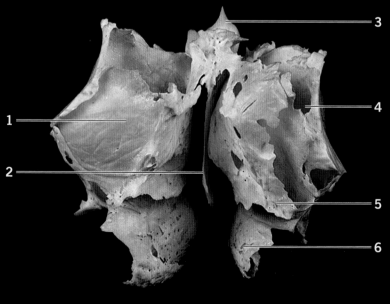

FIGURE 3.18 Ethmoid, Posterior View

1. Lateral mass
2. Perpendicular plate
3. Crista galli
4. Ethmoidal air cell
5. Superior nasal concha
6. Middle nasal concha

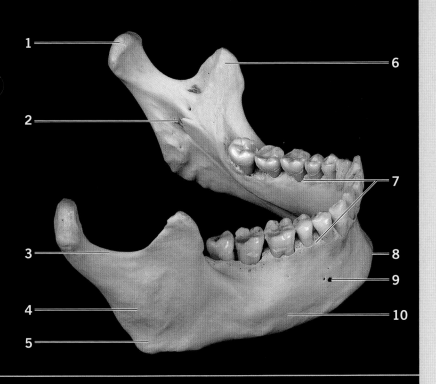

FIGURE 3.19 Mandible, Right Lateral View

1. Mandibular condyle
2. Mandibular foramen
3. Mandibular notch
4. Ramus
5. Mandibular angle
6. Coronoid process
7. Alveolar margin
8. Mental protuberance
9. Mental foramen
10. Body

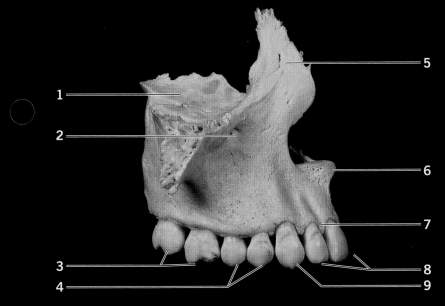

FIGURE 3.20 Right Maxilla, Lateral View

1. Orbital surface
2. Infraorbital foramen
3. Molars
4. Premolars
5. Frontal process
6. Anterior nasal spine
7. Alveolar margin
8. Incisors
9. Canine

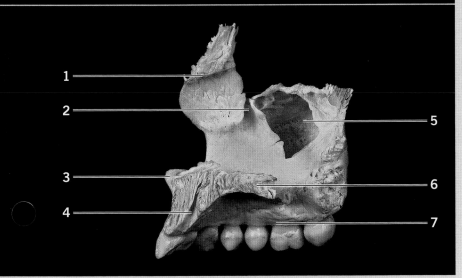

FIGURE 3.21 Right Maxilla, Medial View

1. Frontal process
2. Lacrimal groove
3. Anterior nasal spine
4. Incisive canal
5. Maxillary sinus
6. Palatine process
7. Alveolar margin

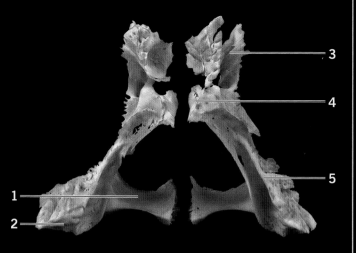

FIGURE 3.22 Palatine Bones, Posterior View

1. Horizontal plate
2. Pyramidal process
3. Orbital process
4. Sphenoidal process
5. Perpendicular plate

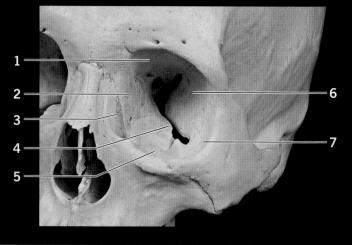

FIGURE 3.23 Bones of the Orbit, Left Eye

1. Frontal bone
2. Ethmoid bone
3. Lacrimal bone
4. Palatine bone
5. Maxilla
6. Sphenoid bone
7. Zygomatic bone

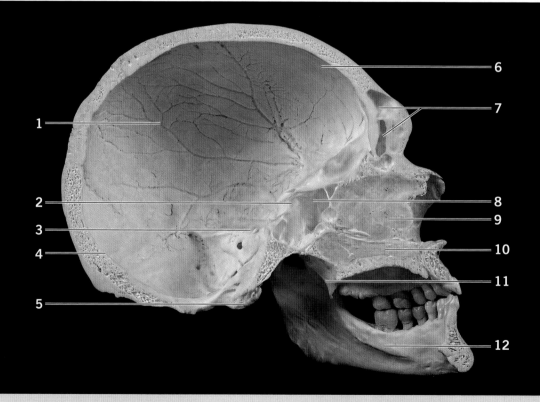

FIGURE 3.24 Bisected Skull

1. Parietal bone
2. Sella turcica
3. Temporal bone
4. Occipital bone
5. Occipital condyle
6. Frontal bone
7. Frontal sinus
8. Sphenoid sinus
9. Ethmoid
10. Vomer
11. Pterygoid process
12. Mandible

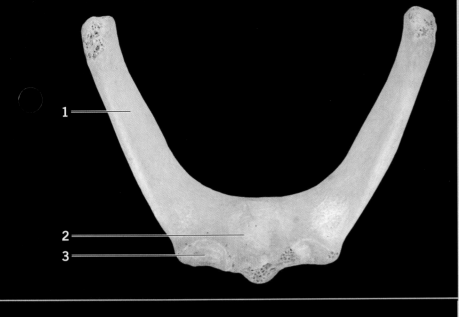

FIGURE 3.25 Hyoid Bone

1. Greater horn
2. Body
3. Lesser horn

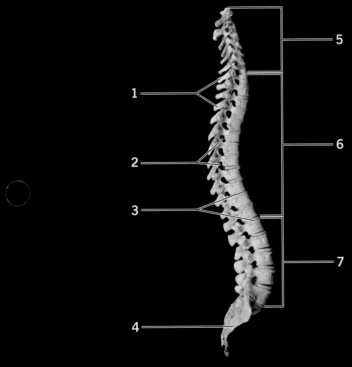

FIGURE 3.26 Vertebral Column, Right Lateral View

1. Spinous processes
2. Intervertebral foramina
3. Intervertebral discs
4. Sacrum and sacral curvature
5. Cervical vertebrae and curvature
6. Thoracic vertebrae and curvature
7. Lumbar vertebrae and curvature

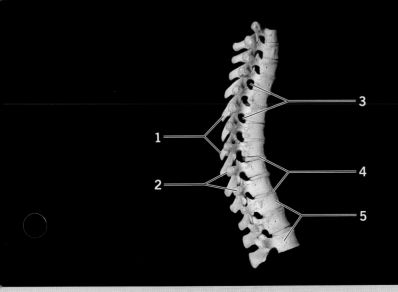

FIGURE 3.27 Thoracic Vertebrae

1. Spinous processes
2. Transverse processes
3. Intervertebral foramina
4. Intervertebral discs
5. Vertebral bodies

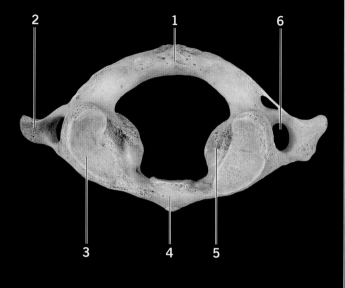

FIGURE 3.28 Atlas, Superior View

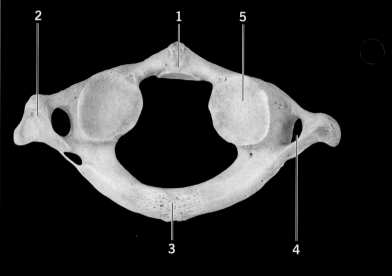

FIGURE 3.29 Atlas, Inferior View

1. Posterior arch
2. Transverse process
3. Superior articular facet
4. Anterior arch
5. Lateral mass
6. Transverse foramen

1. Anterior arch
2. Transverse process
3. Posterior arch
4. Transverse foramen
5. Inferior articular facet

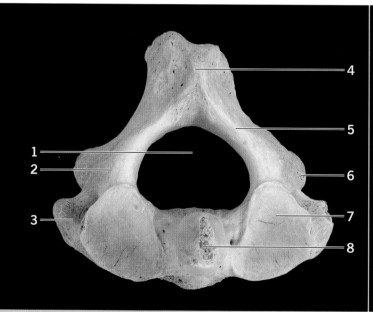

FIGURE 3.30 Axis, Superior View

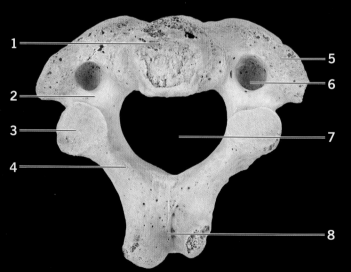

FIGURE 3.31 Axis, Inferior View

1. Vertebral foramen
2. Pedicle
3. Transverse process
4. Spinous process
5. Lamina
6. Inferior articular process
7. Superior articular facet
8. Dens (odontoid process)

1. Body
2. Pedicle
3. Inferior articular facet
4. Lamina
5. Transverse process
6. Transverse foramen
7. Vertebral foramen
8. Spinous process

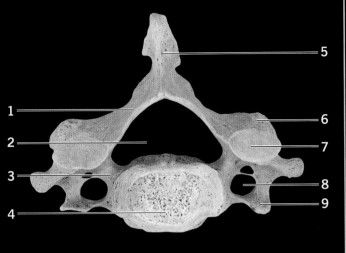

FIGURE 3.32 Cervical Vertebra, Superior View

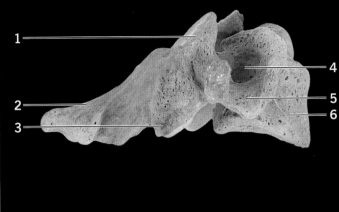

FIGURE 3.33 Cervical Vertebra, Lateral View

1. Lamina
2. Vertebral foramen
3. Pedicle
4. Body
5. Spinous process

6. Inferior articular process
7. Superior articular facet
8. Transverse foramen
9. Transverse process

1. Superior articular process
2. Spinous process
3. Inferior articular process

4. Transverse foramen
5. Transverse process
6. Body

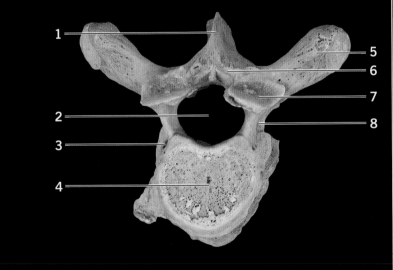

FIGURE 3.34 Thoracic Vertebra, Superior View

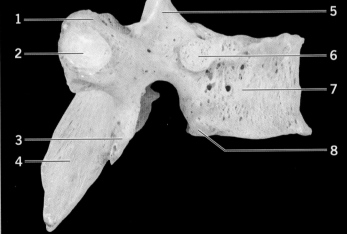

FIGURE 3.35 Thoracic Vertebra, Lateral View

1. Spinous process
2. Vertebral foramen
3. Costal facet
4. Body
5. Transverse process

6. Lamina
7. Superior articular facet
8. Pedicle

1. Transverse process
2. Transverse costal facet
3. Inferior articular process
4. Spinous process

5. Superior articular process and facet
6. Costal facet
7. Body
8. Inferior costal facet

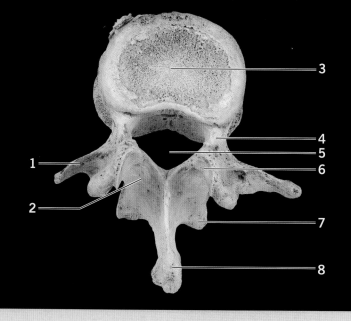

FIGURE 3.36 Lumbar Vertebra, Superior View

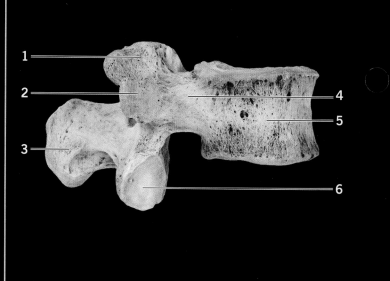

FIGURE 3.37 Lumbar Vertebra, Superior View

1. Transverse process
2. Lamina
3. Body
4. Pedicle
5. Vertebral foramen
6. Superior articular facet
7. Inferior articular process
8. Spinous process

1. Superior articular process
2. Transverse process
3. Spinous process
4. Pedicle
5. Body
6. Inferior articular facet

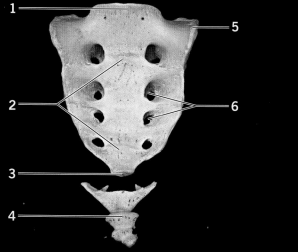

FIGURE 3.38 Sacrum and Coccyx, Anterior View

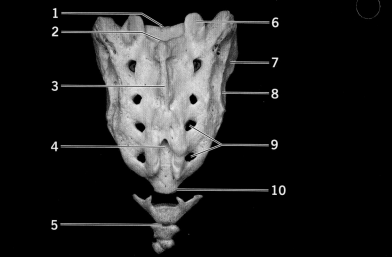

FIGURE 3.39 Sacrum and Coccyx, Posterior View

1. Sacral promontory
2. Transverse lines
3. Apex
4. Coccyx
5. Ala
6. Anterior sacral foramina

1. Body
2. Sacral canal
3. Median sacral crest
4. Sacral hiatus
5. Coccyx
6. Superior articular facet
7. Auricular surface
8. Sacral tuberosity
9. Sacral foramina
10. Apex

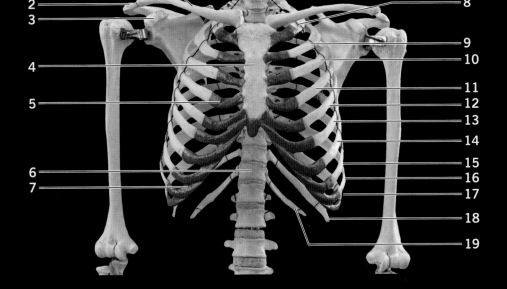

FIGURE 3.40 Thoracic Cage, Anterior View

1. Clavicle
2. Acromion
3. Coracoid process
4. Sternum
5. Costal cartilage of true rib
6. Thoracic vertebra 12
7. Costal cartilage of false rib
8. Rib 1
9. Rib 2
10. Rib 3
11. Rib 4
12. Rib 5
13. Rib 6
14. Rib 7
15. Rib 8
16. Rib 9
17. Rib 10
18. Rib 11
19. Rib 12

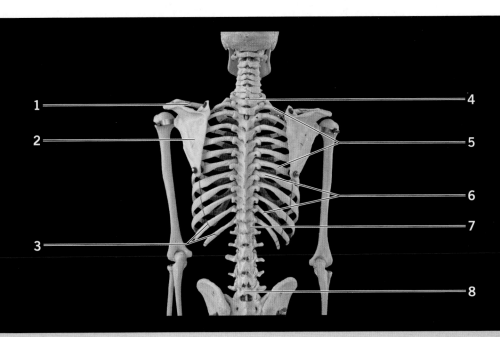

FIGURE 3.41 Thoracic Cage, Posterior View

1. Clavicle
2. Scapula
3. Floating ribs
4. Thoracic vertebra 1
5. True ribs
6. False ribs
7. Thoracic vertebra 12
8. Lumbar vertebra 5

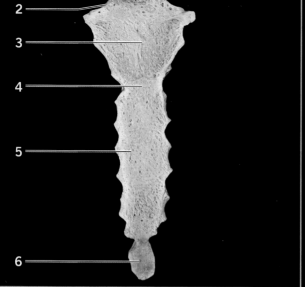

FIGURE 3.42 Sternum, Anterior View

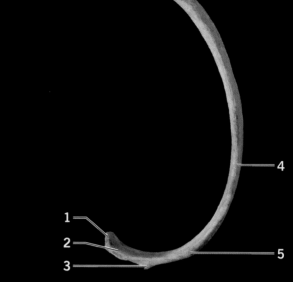

FIGURE 3.43 Right Rib, Superior View

1. Jugular notch	4. Sternal angle
2. Clavicular notch	5. Body
3. Manubrium	6. Xiphoid process

1. Head	4. Body
2. Neck	5. Costal angle
3. Articular tubercle and facet	

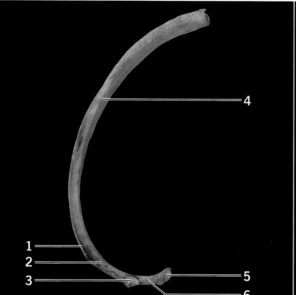

FIGURE 3.44 Right Rib, Inferior View

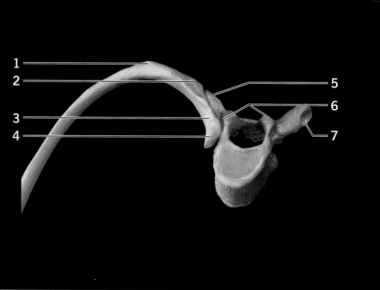

FIGURE 3.45 Articulated Thoracic Vertebra with Right Rib

1. Costal groove	4. Body
2. Costal angle	5. Head
3. Articular tubercle and facet	6. Neck

1. Angle of rib	6. Superior articular processes
2. Tubercle of rib	
3. Neck of rib	7. Transverse costal facet
4. Head of rib	
5. Transverse process	

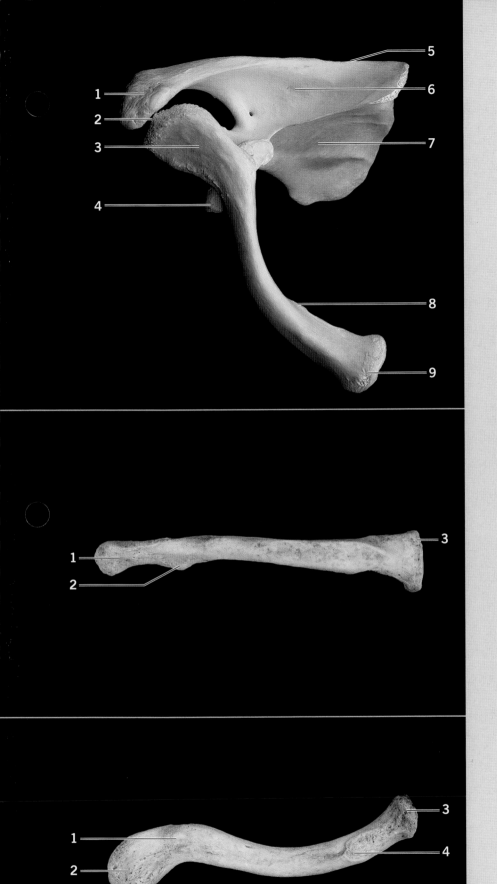

FIGURE 3.46 Articulated Right Clavicle and Scapula, Superior View

1. Acromion
2. Acromioclavicular joint
3. Acromial end of clavicle
4. Coracoid process
5. Spine
6. Supraspinous fossa
7. Subscapular fossa
8. Shaft of clavicle
9. Sternal end of clavicle

FIGURE 3.47 Right Clavicle, Anterior View

1. Acromial end
2. Conoid tubercle
3. Sternal end

FIGURE 3.48 Right Clavicle, Inferior View

1. Conoid tubercle
2. Acromial end
3. Sternal end
4. Costal tuberosity

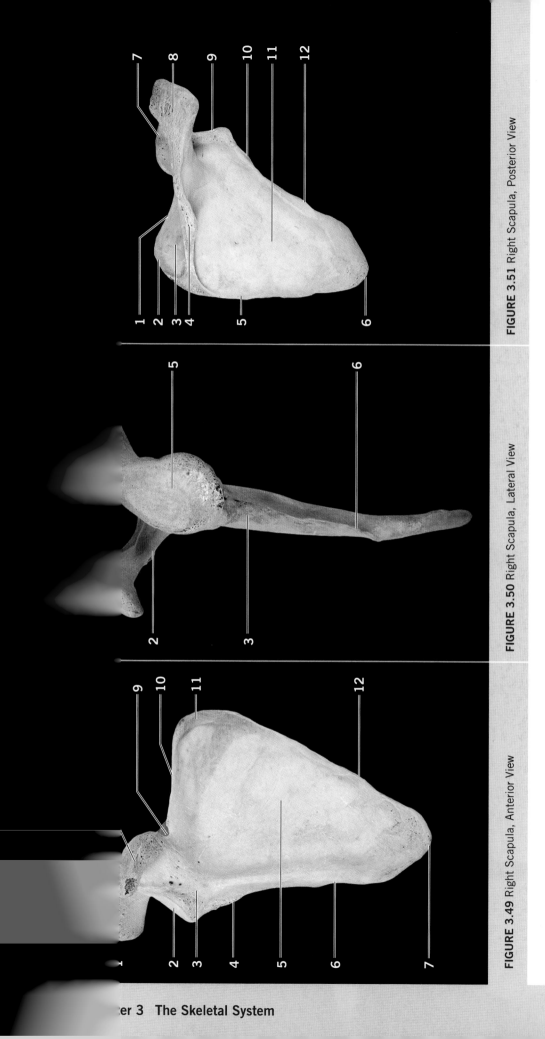

FIGURE 3.49 Right Scapula, Anterior View

1. Acromion
2. Glenoid fossa (cavity)
3. Lateral angle and neck
4. Infraglenoid tubercle
5. Subscapular fossa
6. Lateral (axillary) border
7. Inferior angle
8. Coracoid process
9. Suprascapular notch
10. Superior border
11. Superior angle
12. Medial (vertebral) border

FIGURE 3.50 Right Scapula, Lateral View

1. Acromion
2. Spine
3. Infraglenoid tubercle
4. Coracoid process
5. Glenoid fossa (cavity)
6. Lateral (axillary) border

FIGURE 3.51 Right Scapula, Posterior View

1. Superior border
2. Superior angle
3. Supraspinous fossa
4. Spine
5. Medial (vertebral) border
6. Inferior angle
7. Coracoid process
8. Acromion
9. Glenoid fossa (cavity)
10. Infraglenoid tubercle
11. Infraspinous fossa
12. Lateral (axillary) border

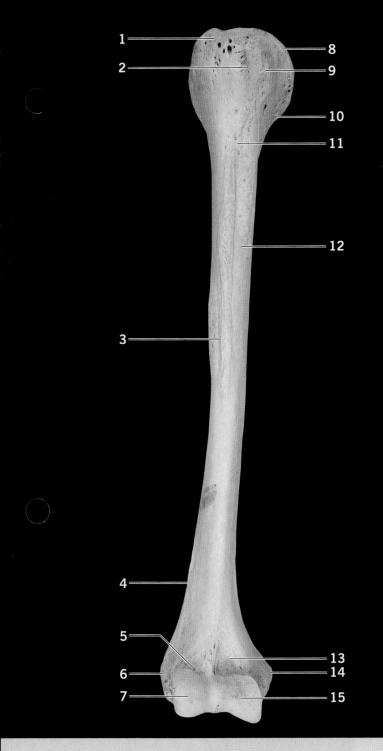

FIGURE 3.52 Right Humerus, Anterior View

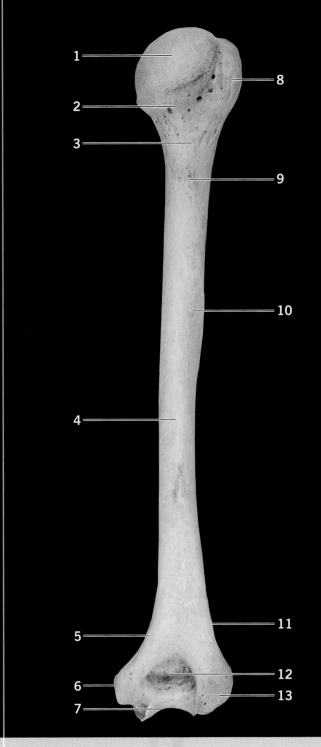

FIGURE 3.53 Right Humerus, Posterior View

1. Greater tubercle
2. Intertubercular sulcus (groove)
3. Deltoid tuberosity
4. Lateral supracondylar ridge
5. Radial fossa
6. Lateral epicondyle
7. Capitulum

8. Head
9. Lesser tubercle
10. Anatomical neck
11. Surgical neck
12. Shaft
13. Coronoid fossa
14. Medial epicondyle
15. Trochlea

1. Head
2. Anatomical neck
3. Surgical neck
4. Radial groove
5. Medial supracondylar ridge
6. Medial epicondyle
7. Trochlea

8. Greater tubercle
9. Shaft
10. Deltoid tuberosity
11. Lateral supracondylar ridge
12. Olecranon fossa
13. Lateral epicondyle

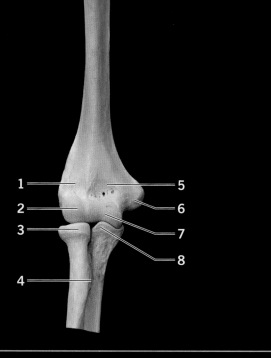

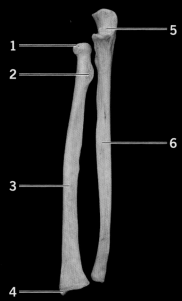

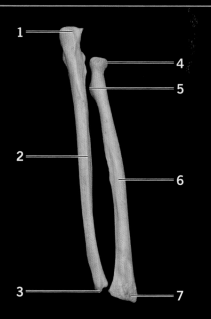

FIGURE 3.54 Articulated Humerus, Ulna, and Radius, Anterior View

1. Radial fossa
2. Capitulum of humerus
3. Head of radius
4. Radial tuberosity
5. Coronoid fossa
6. Medial epicondyle of humerus
7. Trochlea of humerus
8. Coronoid process of ulna

FIGURE 3.55 Right Radius and Ulna, Anterior View

1. Head
2. Radial tuberosity
3. Radius
4. Styloid process of radius
5. Trochlear notch
6. Ulna

FIGURE 3.56 Right Radius and Ulna, Posterior View

1. Olecranon process
2. Ulna
3. Styloid process of ulna
4. Head
5. Radial tuberosity
6. Radius
7. Styloid process of radius

Appendicular Skeleton

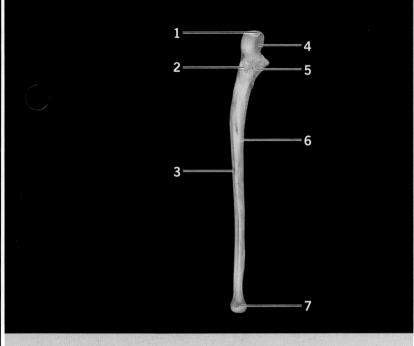

FIGURE 3.57 Right Ulna, Anterior View

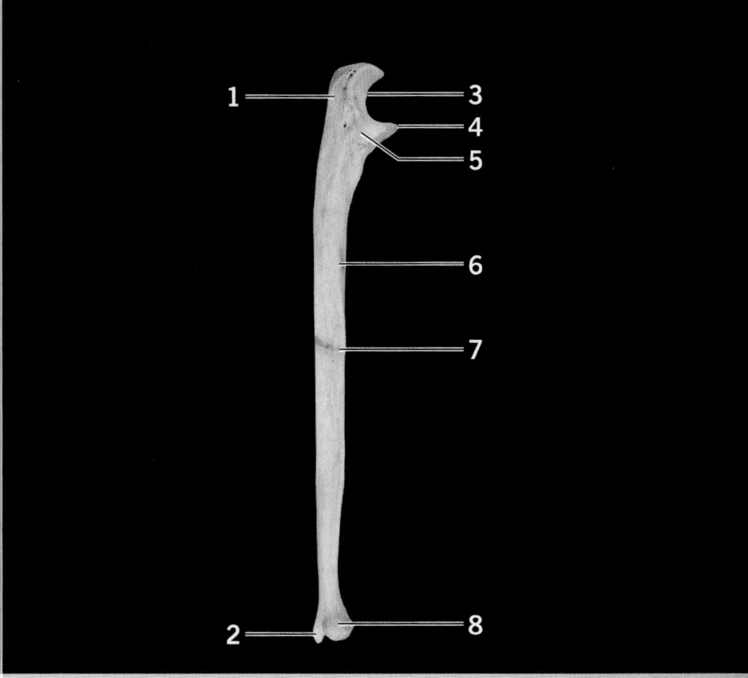

FIGURE 3.58 Right Ulna, Lateral View

1. Olecranon process
2. Radial notch
3. Interosseous border (crest)
4. Trochlear notch
5. Coronoid process
6. Shaft
7. Head

1. Olecranon process
2. Styloid process
3. Trochlear notch
4. Coronoid process
5. Radial notch
6. Shaft
7. Interosseous border (crest)
8. Head

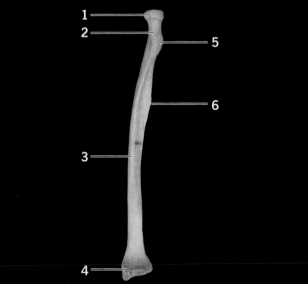

FIGURE 3.59 Right Radius, Anterior View

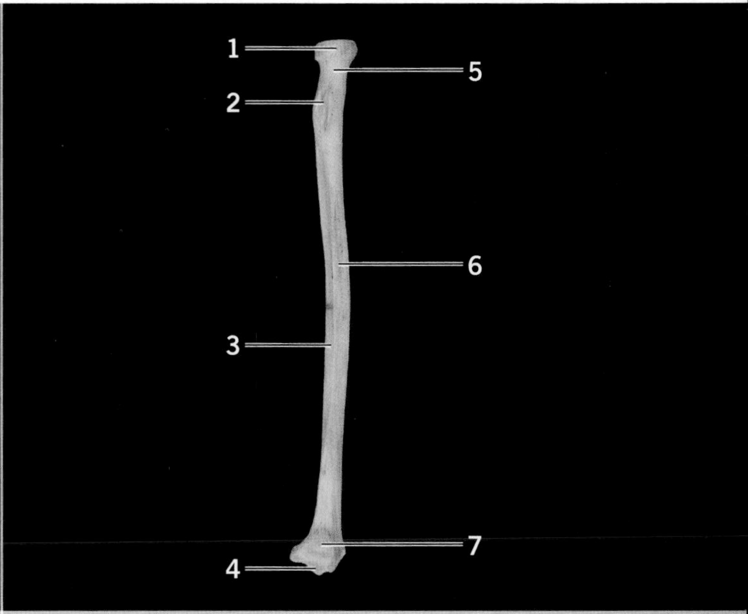

FIGURE 3.60 Right Radius, Medial View

1. Head
2. Neck
3. Shaft
4. Styloid process
5. Radial tuberosity
6. Interosseous border (crest)

1. Head
2. Radial tuberosity
3. Interosseous border (crest)
4. Styloid process
5. Neck
6. Shaft
7. Ulnar notch

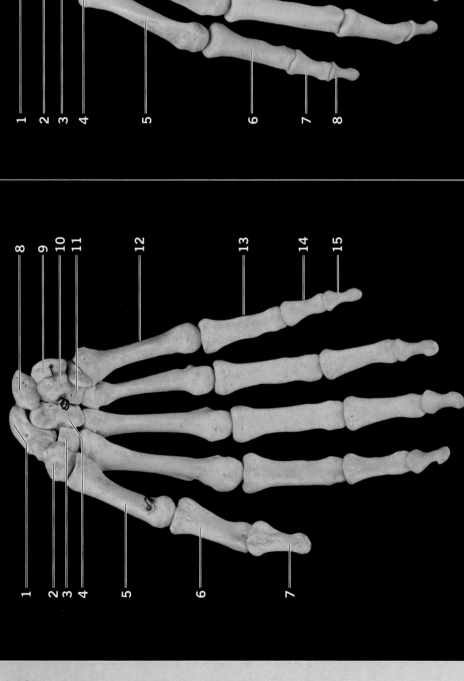

FIGURE 3.61 Bones of the Right Hand, Anterior View

1. Scaphoid
2. Trapezium
3. Trapezoid
4. Capitate
5. Metacarpal 1
6. Proximal phalanx 1
7. Distal phalanx 1
8. Lunate

9. Triquetrum
10. Pisiform
11. Hamate
12. Metacarpal 5
13. Proximal phalanx 5
14. Middle phalanx 5
15. Distal phalanx 5

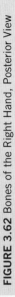

FIGURE 3.62 Bones of the Right Hand, Posterior View

1. Lunate
2. Triquetral
3. Capitate
4. Hamate
5. Metacarpal 5
6. Proximal phalanx 5
7. Middle phalanx 5

8. Distal phalanx 5
9. Scaphoid
10. Trapezium
11. Trapezoid
12. Metacarpal 1
13. Proximal phalanx 1
14. Distal phalanx 1

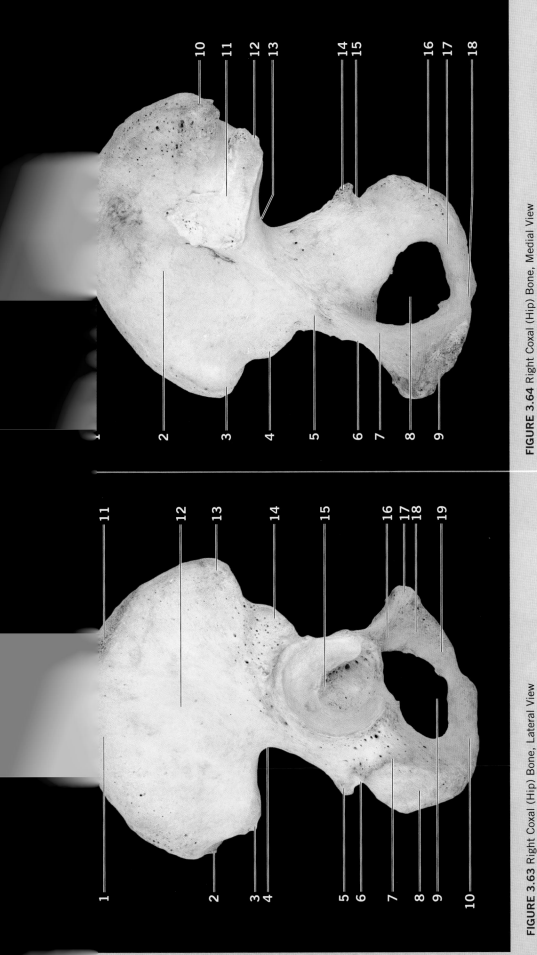

FIGURE 3.63 Right Coxal (Hip) Bone, Lateral View

1. Ilium
2. Posterior superior iliac spine
3. Posterior inferior iliac spine
4. Greater sciatic notch
5. Ischial spine
6. Lesser sciatic notch
7. Ischium
8. Ischial tuberosity
9. Obturator foramen
10. Ischial ramus
11. Iliac crest
12. Ala of ilium
13. Anterior superior iliac spine
14. Anterior inferior iliac spine
15. Acetabulum
16. Superior ramus of pubis
17. Pubic tubercle
18. Pubis
19. Inferior ramus of pubis

FIGURE 3.64 Right Coxal (Hip) Bone, Medial View

1. Iliac crest
2. Iliac fossa
3. Anterior superior iliac spine
4. Anterior inferior iliac spine
5. Arcuate line
6. Pectineal line
7. Superior ramus of pubis
8. Obturator foramen
9. Pubic symphyseal fossa
10. Posterior superior iliac spine
11. Auricular surface of ilium
12. Posterior inferior iliac spine
13. Greater sciatic notch
14. Ischial spine
15. Lesser sciatic notch
16. Ischial tuberosity
17. Ischial ramus
18. Inferior ramus of pubis

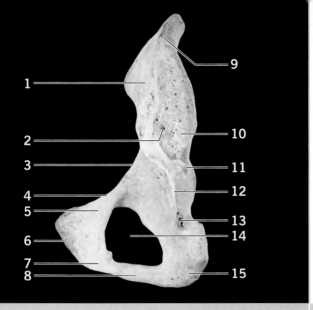

FIGURE 3.65 Right Coxal (Hip) Bone, Posteromedial View

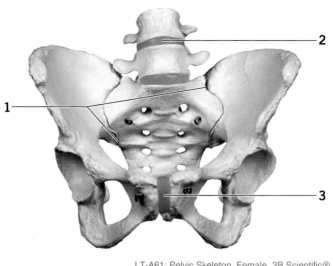

LT-A61: Pelvic Skeleton, Female, 3B Scientific®

FIGURE 3.66 Female Pelvis, Anterior View

1. Iliac fossa
2. Auricular surface of ilium
3. Arcuate line
4. Pectineal line
5. Superior ramus of pubis
6. Pubic symphyseal fossa
7. Inferior ramus of pubis
8. Ischial ramus
9. Iliac crest

10. Posterior superior iliac spine
11. Posterior inferior iliac spine
12. Greater sciatic notch
13. Ischial spine
14. Obturator foramen
15. Ischial tuberosity

1. Sacroiliac joints
2. Intervertebral disc
3. Pubic symphysis

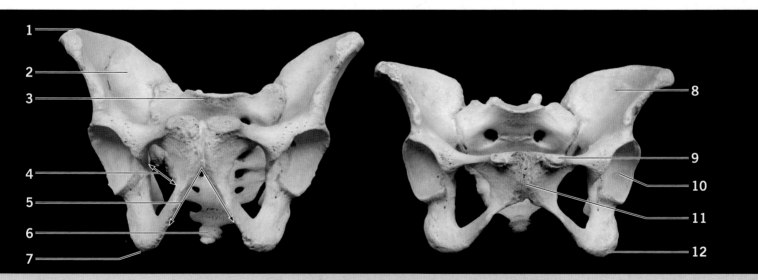

FIGURE 3.67 Male (Left) and Female (Right) Pelves, Anterior View

1. Iliac crest
2. Iliac fossa
3. Sacrum
4. Obturator foramen

5. Pubic angle in male more acute
6. Coccyx in male more ventral

7. Ischial tuberosity
8. Ilium of female flared
9. Pubic tubercle
10. Acetabulum

11. Pubic symphysis
12. Ischial tuberosities farther apart in female

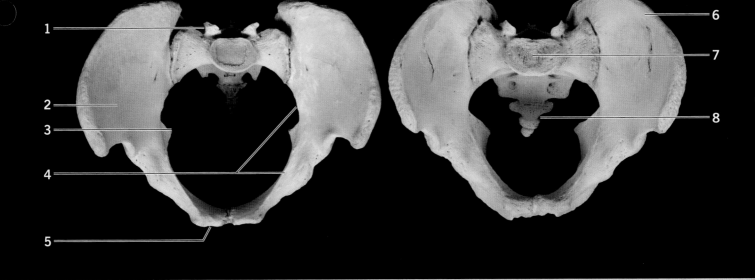

FIGURE 3.68 Female (Left) and Male (Right) Pelves, Superior View

1. Superior articular process
2. Iliac fossa
3. Ischial spine
4. Pelvic brim
5. Pubic crest
6. Iliac crest
7. Body of sacrum
8. Coccyx

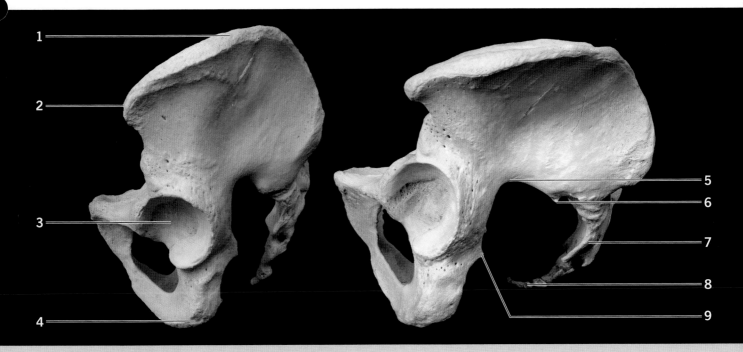

FIGURE 3.69 Male (Left) and Female (Right) Pelves, Lateral View

1. Iliac crest
2. Anterior superior iliac spine
3. Acetabulum
4. Ischial tuberosity
5. Greater sciatic notch
6. Posterior inferior iliac spine
7. Sacrum
8. Coccyx
9. Ischial spine

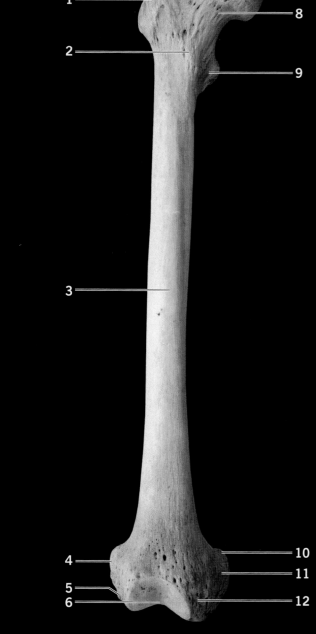

FIGURE 3.70 Right Femur, Anterior View

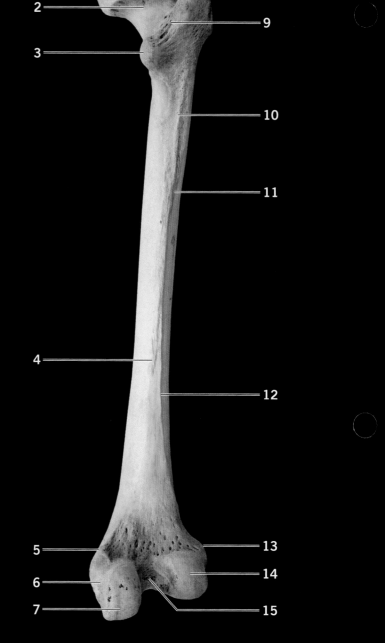

FIGURE 3.71 Right Femur, Posterior View

1. Greater trochanter
2. Intertrochanteric line
3. Shaft
4. Lateral epicondyle
5. Lateral condyle
6. Patellar surface
7. Head
8. Neck
9. Lesser trochanter
10. Adductor tubercle
11. Medial epicondyle
12. Medial condyle

1. Head
2. Neck
3. Lesser trochanter
4. Medial supracondylar line
5. Adductor tubercle
6. Medial epicondyle
7. Medial condyle
8. Greater trochanter
9. Intertrochanteric crest
10. Gluteal tuberosity
11. Linea aspera
12. Lateral supracondylar line
13. Lateral epicondyle
14. Lateral condyle
15. Intercondylar fossa

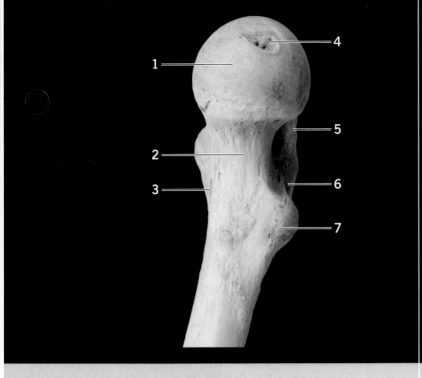

FIGURE 3.72 Proximal End of Right Femur, Medial View

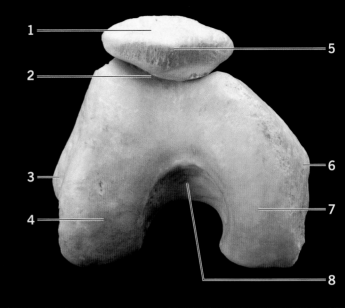

FIGURE 3.73 Right Femur and Patella, Inferior View

1. Head
2. Neck
3. Intertrochanteric line
4. Fovea capitis
5. Greater trochanter
6. Intertrochanteric crest
7. Lesser trochanter

1. Patella
2. Patellofemoral joint
3. Lateral epicondyle
4. Lateral condyle
5. Apex
6. Medial epicondyle
7. Medial condyle
8. Intercondylar fossa

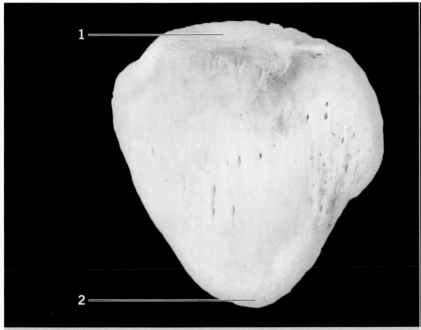

FIGURE 3.74 Right Patella, Anterior View

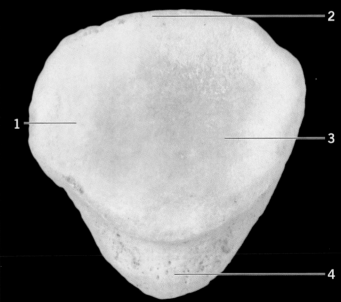

FIGURE 3.75 Right Patella, Posterior View

1. Base
2. Apex

1. Facet for medial condyle of femur
2. Base
3. Facet for lateral condyle of femur
4. Apex

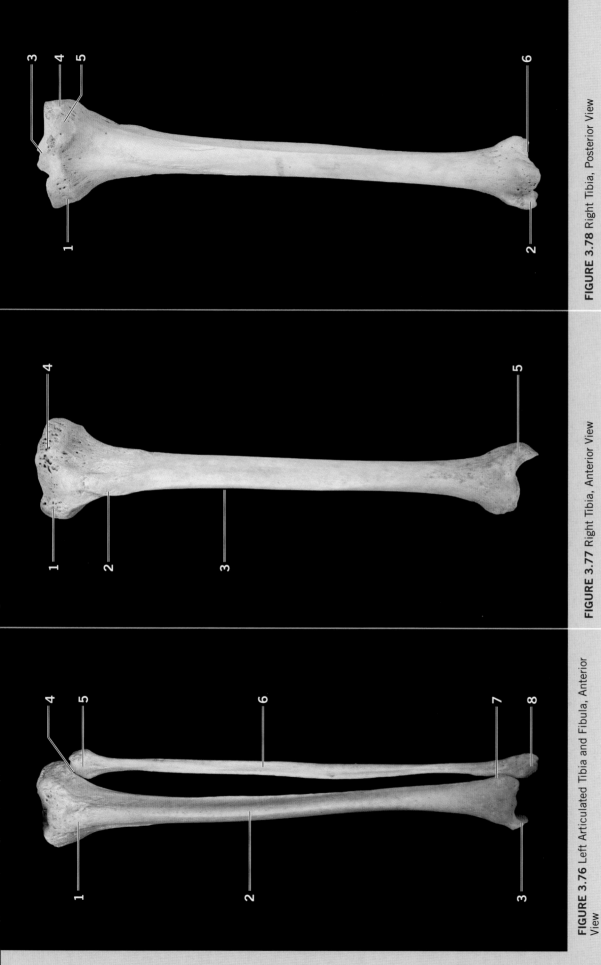

FIGURE 3.76 Left Articulated Tibia and Fibula, Anterior View

1. Tibial tuberosity
2. Shaft of tibia
3. Medial malleolus
4. Proximal tibiofibular joint
5. Head of fibula
6. Shaft of fibula
7. Distal tibiofibular joint
8. Lateral malleolus

FIGURE 3.77 Right Tibia, Anterior View

1. Lateral condyle
2. Tibial tuberosity
3. Anterior crest
4. Medial condyle
5. Medial malleolus

FIGURE 3.78 Right Tibia, Posterior View

1. Medial condyle
2. Medial malleolus
3. Intercondylar eminence
4. Lateral condyle
5. Articular facet for head of fibula
6. Fibular notch

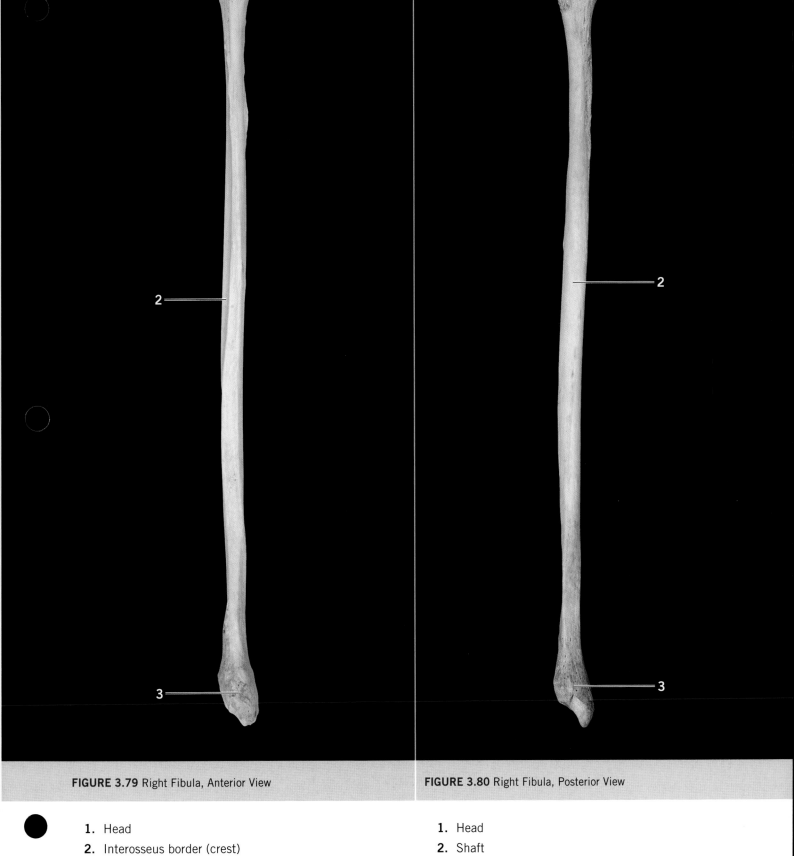

FIGURE 3.79 Right Fibula, Anterior View

FIGURE 3.80 Right Fibula, Posterior View

1. Head
2. Interosseus border (crest)
3. Lateral malleolus

1. Head
2. Shaft
3. Lateral malleolus

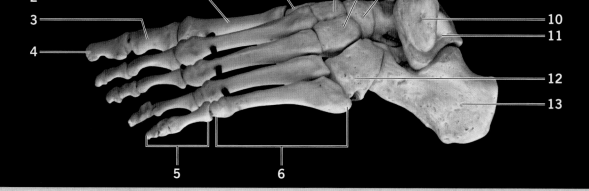

FIGURE 3.81 Bones of the Left Foot and Distal Fibula, Lateral View

1. Medial cuneiform
2. Metatarsal 1
3. Proximal phalanx 1
4. Distal phalanx 1

5. Phalanges
6. Metatarsals
7. Intermediate cuneiform

8. Lateral cuneiform
9. Navicular
10. Fibula
11. Talus

12. Cuboid
13. Calcaneus

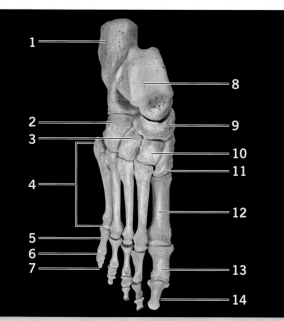

FIGURE 3.82 Bones of the Right Foot, Superior View

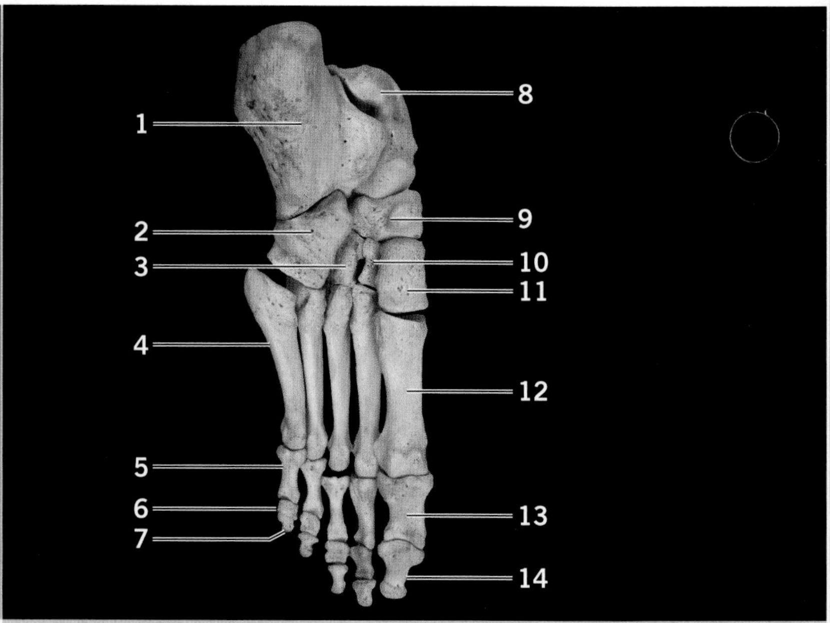

1. Calcaneus
2. Cuboid
3. Lateral cuneiform
4. Metatarsals
5. Proximal phalanx 5
6. Middle phalanx 5
7. Distal phalanx 5
8. Talus

9. Navicular
10. Intermediate cuneiform
11. Medial cuneiform
12. Metatarsal 1
13. Proximal phalanx 1
14. Distal phalanx 1

FIGURE 3.83 Bones of the Left Foot, Inferior View

1. Calcaneus
2. Cuboid
3. Lateral cuneiform
4. Metatarsal 5
5. Proximal phalanx 5
6. Middle phalanx 5
7. Distal phalanx 5
8. Talus

9. Navicular
10. Intermediate cuneiform
11. Medial cuneiform
12. Metatarsal 1
13. Proximal phalanx 1
14. Distal phalanx 1

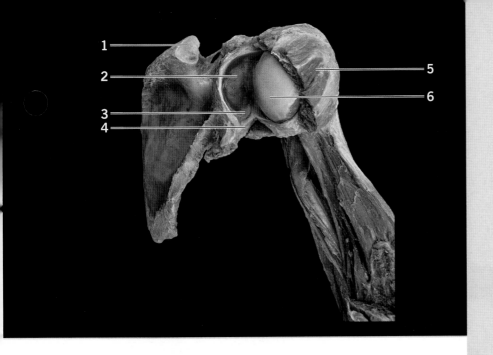

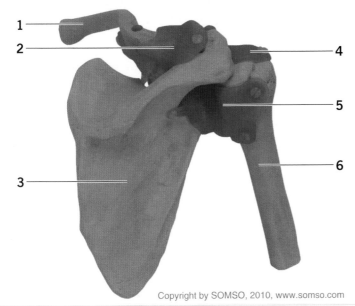

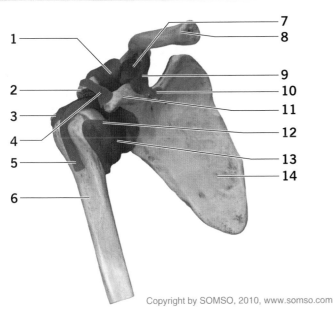

FIGURE 3.84 Shoulder Joint, Opened, Posterior View

1. Acromion (cut)
2. Glenoid cavity
3. Glenoid labrum
4. Articular capsule
5. Muscles of rotator cuff
6. Articular cartilage on head of humerus

FIGURE 3.85 Shoulder Joint, Posterior View

1. Clavicle
2. Acromioclavicular ligament
3. Scapula
4. Coracohumeral ligament
5. Articular capsule
6. Humerus

FIGURE 3.86 Shoulder Joint, Anterior View

1. Acromioclavicular ligament
2. Acromion
3. Tendon of supraspinatus muscle
4. Coracoacromial ligament
5. Tendon of long head of biceps brachii muscle
6. Humerus
7. Trapezoid ligament (part of coracoclavicular ligament)
8. Clavicle
9. Conoid ligament (part of coracoclavicular ligament)
10. Superior transverse scapular ligament
11. Coracoid process
12. Articular capsule
13. Tendon of subscapularis muscle
14. Scapula

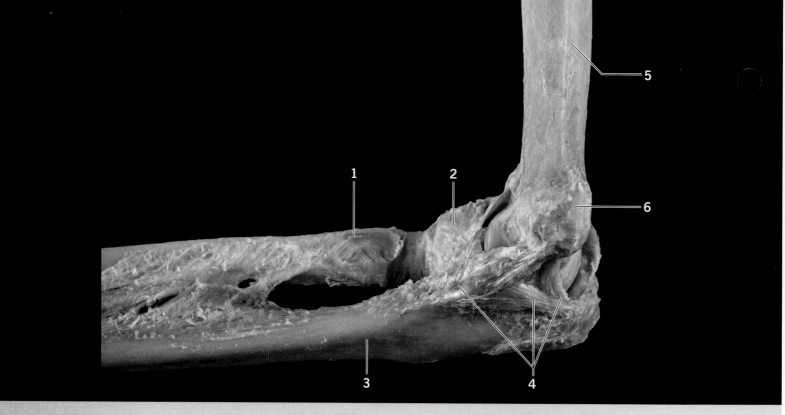

FIGURE 3.87 Right Elbow Joint, Medial View

1. Radius	**3.** Ulna	**5.** Humerus
2. Anular ligament	**4.** Ulnar collateral ligament	**6.** Medial epicondyle of humerus

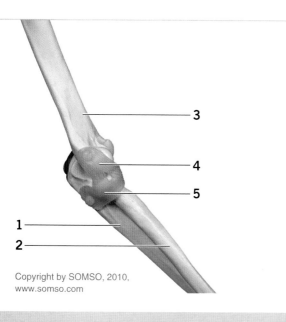

Copyright by SOMSO, 2010,
www.somso.com

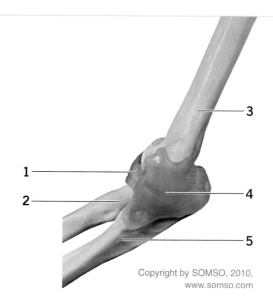

Copyright by SOMSO, 2010,
www.somso.com

FIGURE 3.88 Elbow Joint, Lateral View

1. Ulna	**4.** Radial collateral ligament
2. Radius	**5.** Anular ligament
3. Humerus	

FIGURE 3.89 Elbow Joint, Medial View

1. Anular ligament	**4.** Ulnar collateral ligament
2. Radius	**5.** Ulna
3. Humerus	

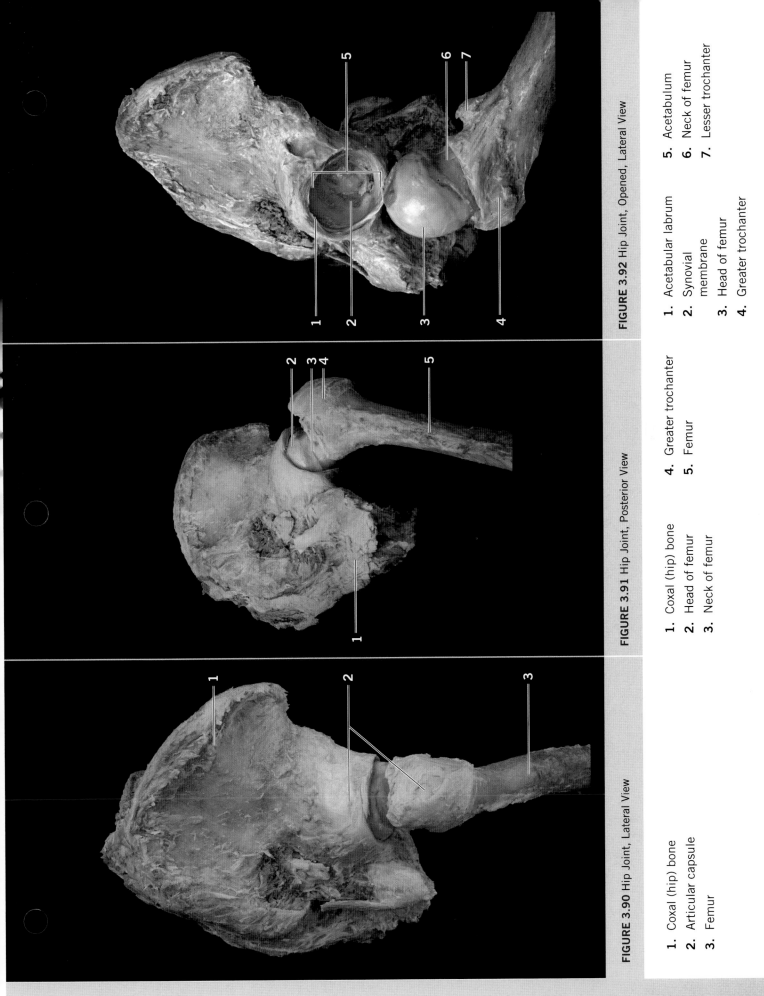

FIGURE 3.90 Hip Joint, Lateral View

1. Coxal (hip) bone
2. Articular capsule
3. Femur

FIGURE 3.91 Hip Joint, Posterior View

1. Coxal (hip) bone
2. Head of femur
3. Neck of femur
4. Greater trochanter
5. Femur

FIGURE 3.92 Hip Joint, Opened, Lateral View

1. Acetabular labrum
2. Synovial membrane
3. Head of femur
4. Greater trochanter
5. Acetabulum
6. Neck of femur
7. Lesser trochanter

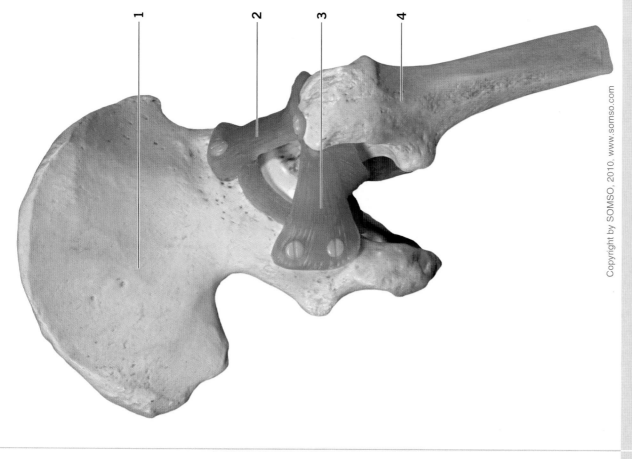
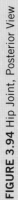

Copyright by SOMSO, 2010, www.somso.com

FIGURE 3.93 Hip Joint, Anterior View

1. Iliofemoral ligament
2. Femur
3. Ilium of coxal bone
4. Pubofemoral ligament

Copyright by SOMSO, 2010, www.somso.com

FIGURE 3.94 Hip Joint, Posterior View

1. Ilium of coxal bone
2. Iliofemoral ligament
3. Ischiofemoral ligament
4. Femur

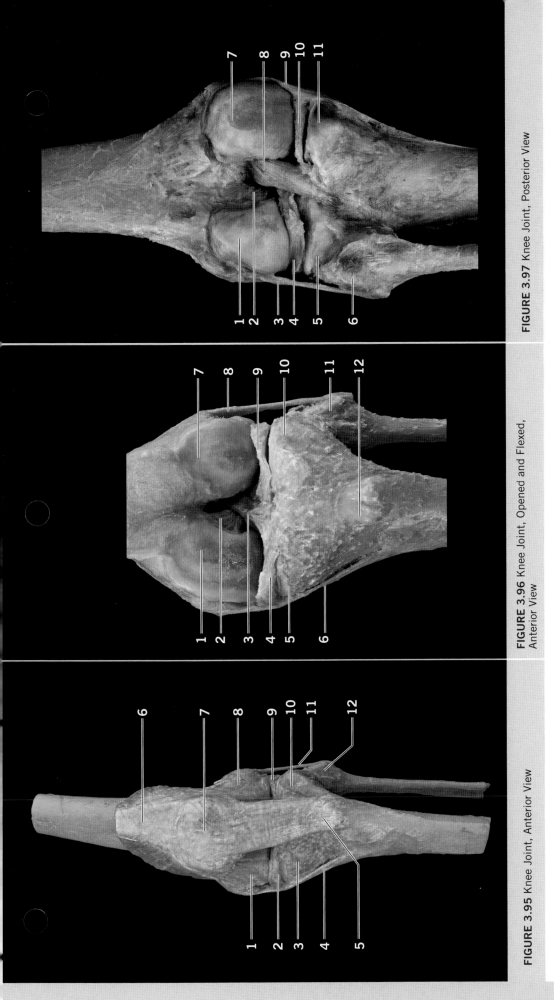

FIGURE 3.95 Knee Joint, Anterior View

1. Medial condyle of femur
2. Medial meniscus
3. Medial condyle of tibia
4. Tibial collateral ligament
5. Patellar ligament
6. Quadriceps tendon
7. Patella
8. Lateral condyle of femur
9. Lateral meniscus
10. Lateral condyle of tibia
11. Fibular collateral ligament
12. Head of fibula

FIGURE 3.96 Knee Joint, Opened and Flexed, Anterior View

1. Medial condyle of femur
2. Posterior cruciate ligament
3. Anterior cruciate ligament
4. Medial meniscus
5. Medial condyle of tibia
6. Tibial collateral ligament
7. Lateral condyle of femur
8. Fibular collateral ligament
9. Lateral meniscus
10. Lateral condyle of tibia
11. Head of fibula
12. Tibial tuberosity

FIGURE 3.97 Knee Joint, Posterior View

1. Lateral condyle of femur
2. Anterior cruciate ligament
3. Fibular collateral ligament
4. Lateral meniscus
5. Lateral condyle of tibia
6. Head of fibula
7. Medial condyle of femur
8. Posterior cruciate ligament
9. Tibial collateral ligament
10. Medial meniscus
11. Medial condyle of tibia

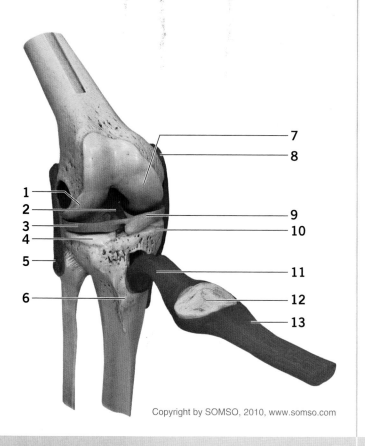

FIGURE 3.98 Knee Joint, Opened and Flexed, Anterior View

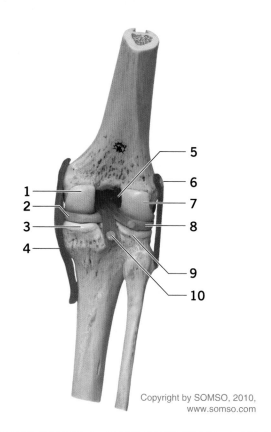

FIGURE 3.99 Knee Joint, Posterior View

1. Articular surface of lateral condyle of femur
2. Anterior cruciate ligament
3. Lateral meniscus
4. Articular surface of lateral condyle of tibia
5. Fibular collateral ligament
6. Tibial tuberosity

7. Articular surface of medial condyle of femur
8. Tibial collateral ligament
9. Medial meniscus
10. Articular surface of medial condyle of tibia
11. Patellar ligament
12. Patella
13. Tendon of quadriceps femoris muscle

1. Articular surface of medial condyle of femur
2. Medial meniscus
3. Articular surface of medial condyle of tibia
4. Tibial collateral ligament
5. Anterior cruciate ligament

6. Fibular collateral ligament
7. Articular surface of lateral condyle of femur
8. Lateral meniscus
9. Articular surface of lateral condyle of tibia
10. Posterior cruciate ligament

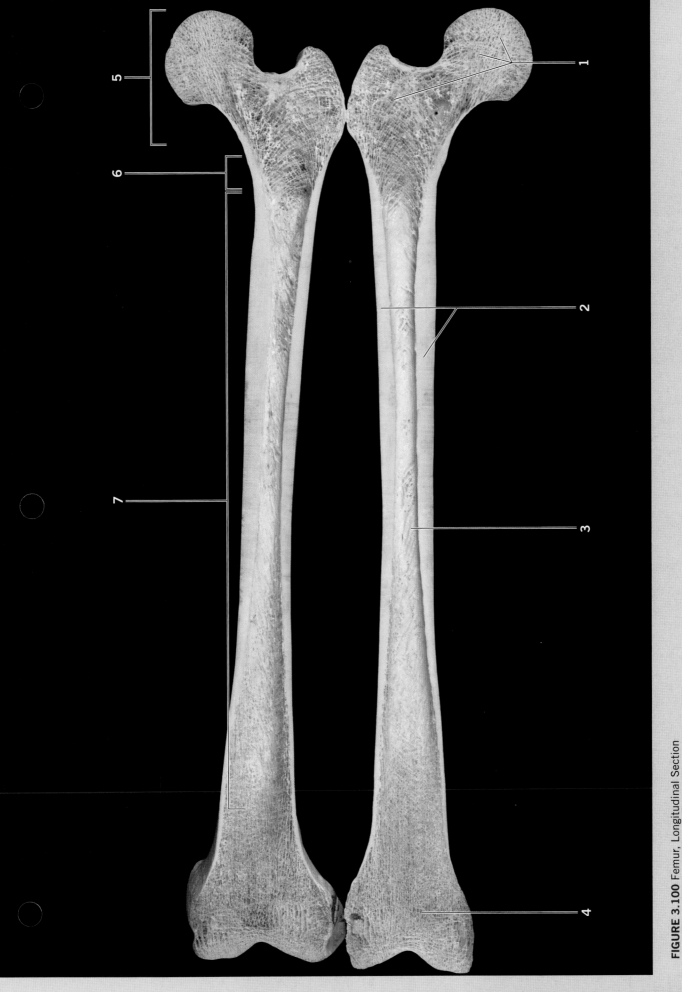

FIGURE 3.100 Femur, Longitudinal Section

1. Spongy bone
2. Compact bone
3. Medullary cavity
4. Epiphyseal line
5. Epiphysis
6. Metaphysis
7. Diaphysis

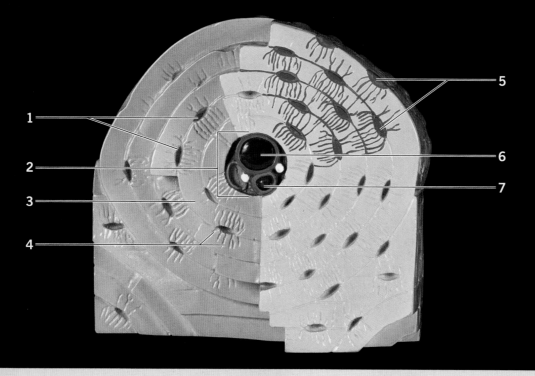

FIGURE 3.101 Osteon Model, Superior View

1. Lacunae
2. Central canal
3. Lamella

4. Canaliculi
5. Osteocytes

6. Vein
7. Artery

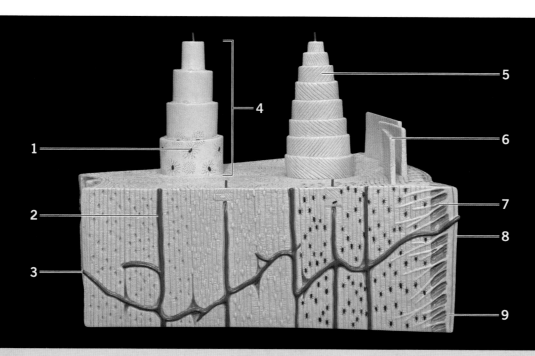

FIGURE 3.102 Osteon Model, Lateral View

1. Osteocyte in lacuna
2. Artery in central canal
3. Artery in perforating canal

4. Osteon
5. Lamella
6. Circumferential lamella

7. Perforating fibers
8. Fibrous layer of periosteum
9. Osteogenic layer of periosteum

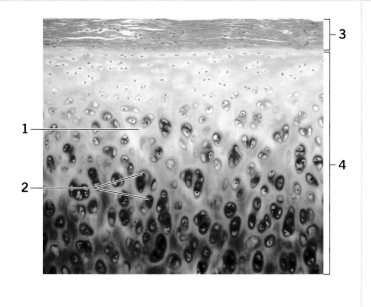

FIGURE 3.103 Hyaline Cartilage, 400×

1. Matrix
2. Chondrocytes in lacunae
3. Perichondrium
4. Hyaline cartilage

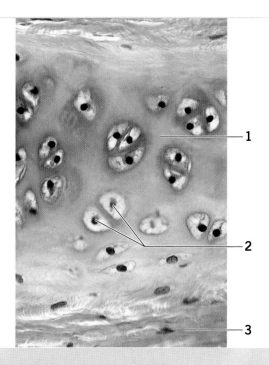

FIGURE 3.104 Hyaline Cartilage, 1000×

1. Matrix
2. Chondrocytes in lacunae
3. Perichondrium

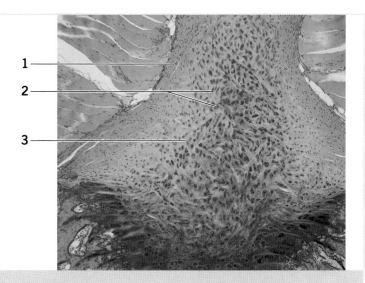

FIGURE 3.105 Fibrocartilage, Pubic Symphysis, 100×

1. Perichondrium
2. Chondrocytes in lacunae
3. Matrix

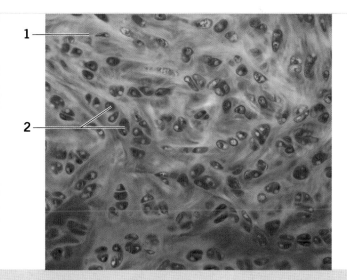

FIGURE 3.106 Fibrocartilage, Pubic Symphysis, 400×

1. Matrix
2. Chondrocytes in lacunae

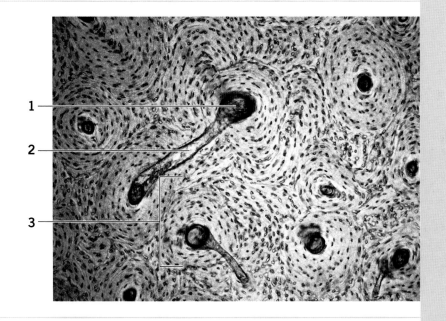

FIGURE 3.107 Compact Bone, Ground, 100×

1. Central canal
2. Perforating canal
3. Osteon (Haversian system)

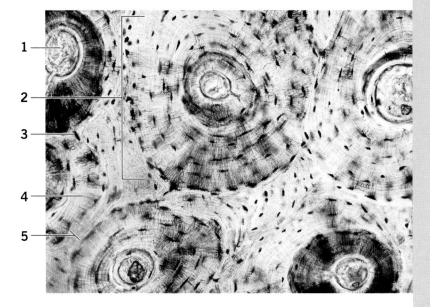

FIGURE 3.108 Compact Bone, Ground, 200×

1. Central canal
2. Osteon
3. Lacuna
4. Canaliculus
5. Lamella

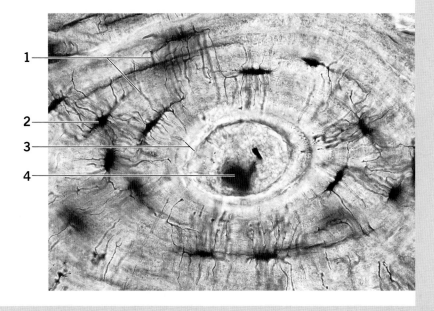

FIGURE 3.109 Compact Bone, Ground, 400×

1. Canaliculi
2. Lacuna
3. Lamella
4. Central canal

Tissues of the Skeletal System

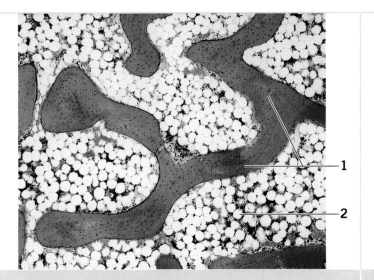

FIGURE 3.110 Spongy Bone, 200×

1. Trabeculae
2. Yellow (fatty) bone marrow

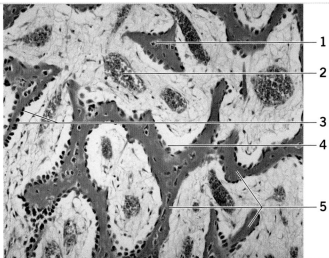

FIGURE 3.111 Spongy Bone, 400×

1. Osteocyte in lacuna   4. Endosteum
2. Blood vessel   5. Trabeculae
3. Osteoblasts

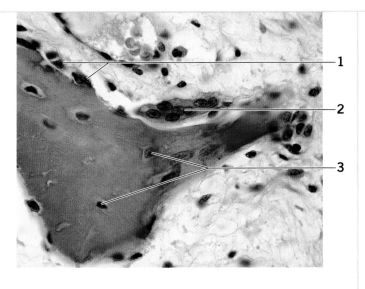

FIGURE 3.112 Spongy Bone, 1000×

1. Osteoblasts   3. Osteocytes in lacunae
2. Osteoclast

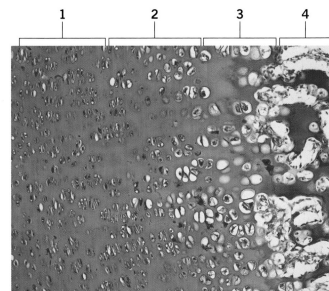

FIGURE 3.113 Epiphyseal Line, Developing Long Bone, 200×

1. Proliferation zone   3. Calcification zone
2. Hypertrophic zone   4. Ossification zone

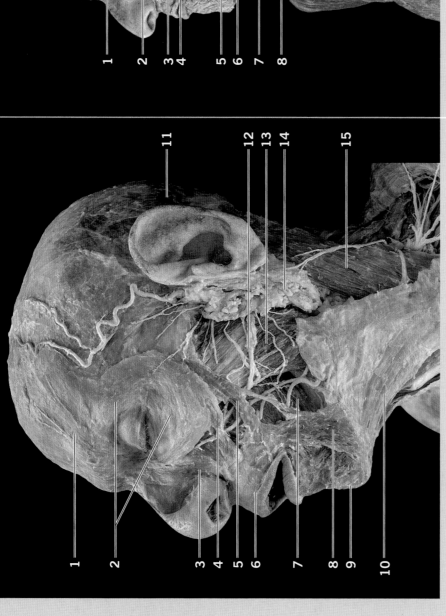

FIGURE 4.1 Superficial Muscles of the Head and Neck, Lateral View

1. Epicranius, frontal belly
2. Orbicularis oculi
3. Levator labii superioris
4. Zygomaticus minor
5. Zygomaticus major
6. Orbicularis oris
7. Buccinator
8. Depressor anguli oris
9. Mentalis
10. Platysma
11. Epicranius, occipital belly
12. *Parotid duct*
13. Masseter
14. *Parotid gland*
15. Sternocleidomastoid

FIGURE 4.2 Intermediate Muscles of the Head and Neck, Lateral View

1. Orbicularis oculi
2. Zygomaticus major
3. *Parotid duct*
4. Masseter
5. Submandibular gland
6. Digastric, anterior belly
7. Omohyoid, superior belly
8. Sternohyoid
9. Temporalis
10. *Parotid gland*
11. Sternocleidomastoid
12. *External jugular vein*
13. Levator scapulae
14. *Clavicle*

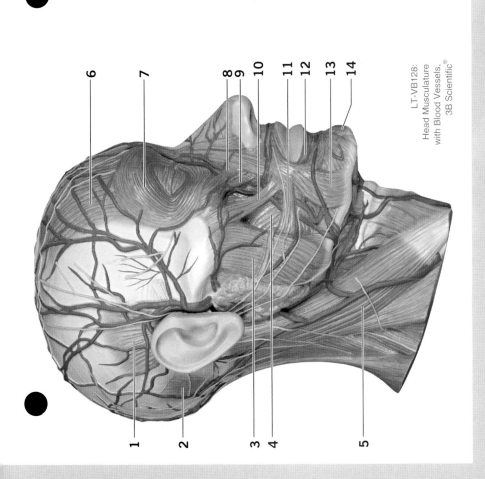

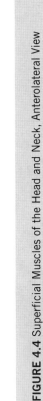

Muscles of the Head and Neck

LT-VB128:
Head Musculature
with Blood Vessels,
3B Scientific®

FIGURE 4.3 Muscles of the Head and Neck, Lateral View

1. Auricularis superior
2. Epicranius, occipital belly
3. Masseter
4. Buccinator
5. Sternocleidomastoid
6. Epicranius, frontal belly
7. Orbicularis oculi
8. Levator labii superioris
9. Zygomaticus minor
10. Zygomaticus major
11. Risorius
12. Orbicularis oris
13. Depressor anguli oris
14. Mentalis

LT-VA16: Life-size
Muscle Torso,
27-part,
3B Scientific®

FIGURE 4.4 Superficial Muscles of the Head and Neck, Anterolateral View

1. Auricularis superior
2. *Parotid duct*
3. Masseter
4. *Parotid gland*
5. Risorius
6. Digastric, posterior belly
7. Digastric, anterior belly
8. Sternocleidomastoid
9. Trapezius
10. Middle scalene
11. Omohyoid, inferior belly
12. Epicranius, frontal belly
13. Orbicularis oculi
14. Zygomaticus minor
15. Zygomaticus major
16. Buccinator
17. Orbicularis oris
18. Depressor anguli oris
19. Mentalis
20. Mylohoid
21. Thyrohyoid
22. Omohyoid, superior belly
23. Sternohyoid

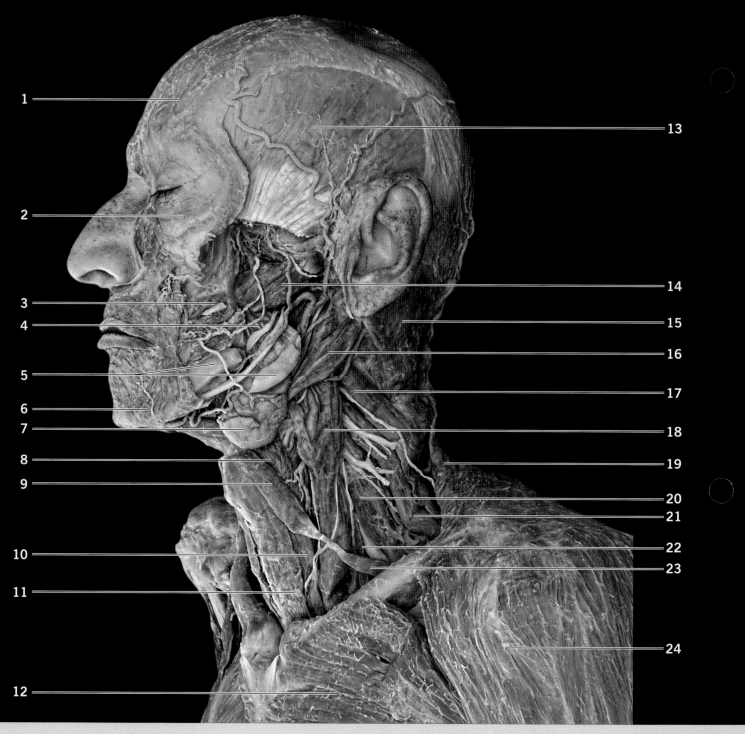

FIGURE 4.5 Deep Muscles of the Head and Neck, Lateral View

1. Epicranius, frontal belly
2. Orbicularis oculi
3. Buccinator
4. Medial pterygoid
5. *Mandible (cut & reflected)*
6. Depressor anguli oris
7. *Submandibular gland*
8. Thyrohyoid
9. Omohyoid, superior belly
10. Sternothyroid
11. Sternohyoid
12. Pectoralis major
13. Temporalis
14. Lateral pterygoid
15. Splenius capitis
16. Digastric, posterior belly
17. Levator scapulae
18. *Internal jugular vein*
19. Trapezius
20. Middle scalene
21. Posterior scalene
22. Anterior scalene
23. Omohyoid, inferior belly
24. Deltoid

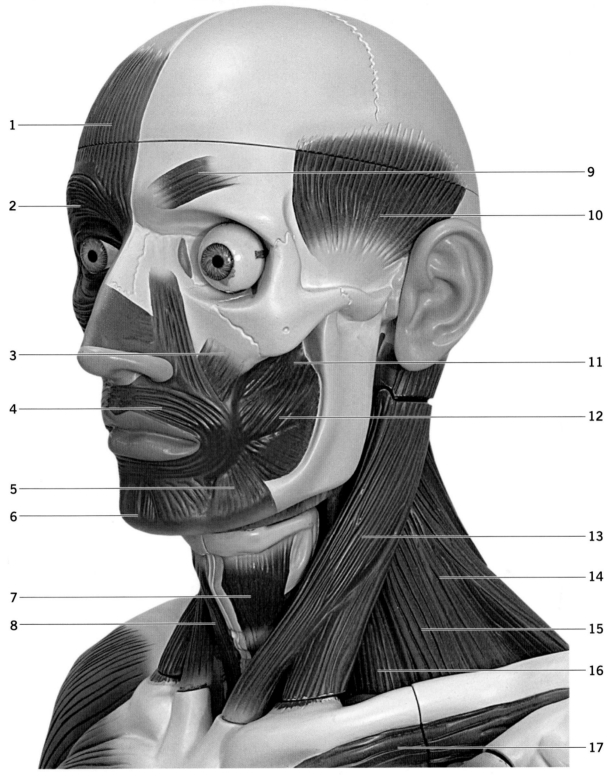

1
2
3
4
5
6
7
8
9
10
11
12
13
14
15
16
17

LT-VA16: Life-size Muscle Torso, 27-part, 3B Scientific®

FIGURE 4.6 Deep Muscles of the Head and Neck, Anterolateral View

1. Epicranius (frontal belly)
2. Orbicularis oculi
3. Levator labii superioris
4. Orbicularis oris
5. Depressor anguli oris
6. Mentalis
7. Thyrohyoid
8. Sternohyoid
9. Corrugator supercilii
10. Temporalis
11. Lateral pterygoid
12. Buccinator
13. Sternocleidomastoid
14. Posterior scalene
15. Middle scalene
16. Anterior scalene
17. Subclavius

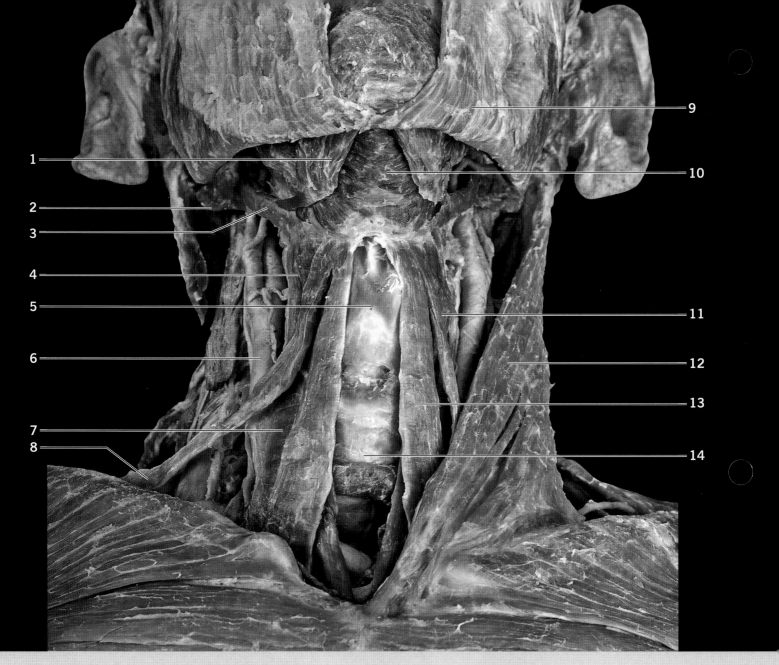

FIGURE 4.7 Muscles of the Neck, Anterior View

1. Digastric, anterior belly
2. Digastric, posterior belly
3. Stylohyoid
4. Thyrohyoid
5. *Thyroid cartilage*
6. *Common carotid artery*
7. Sternothyroid
8. Omohyoid, inferior belly
9. Platysma (reflected)
10. Mylohyoid
11. Omohyoid, superior belly
12. Sternocleidomastoid
13. Sternohyoid
14. *Trachea*

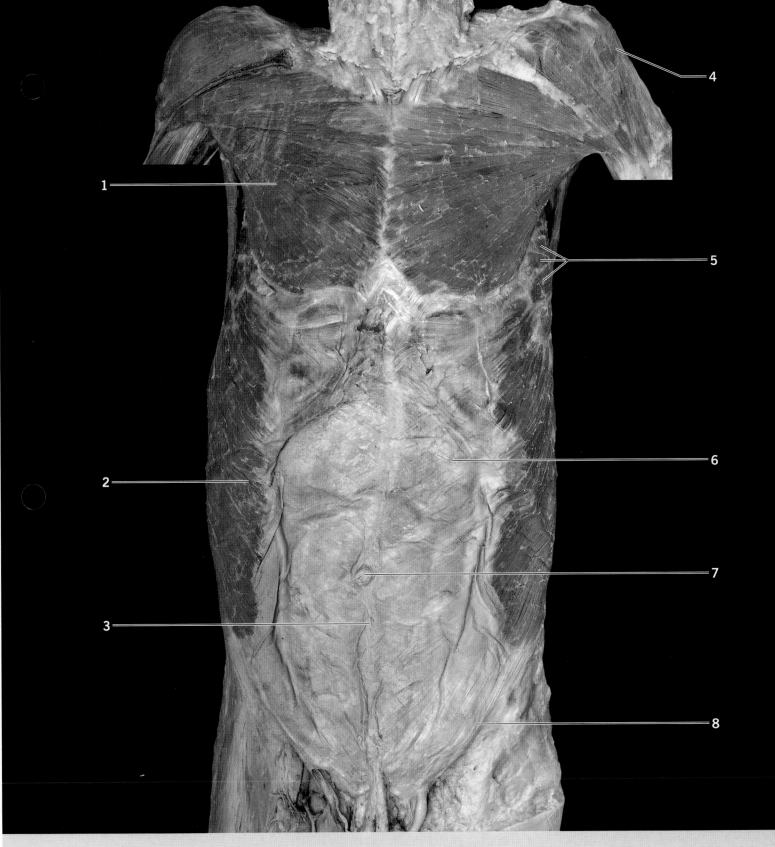

FIGURE 4.8 Superficial Muscles of the Trunk, Anterior View

1. Pectoralis major
2. External oblique
3. *Linea alba*
4. Deltoid
5. Serratus anterior
6. *Rectus sheath, anterior layer*
7. *Umbilicus*
8. *Inguinal ligament*

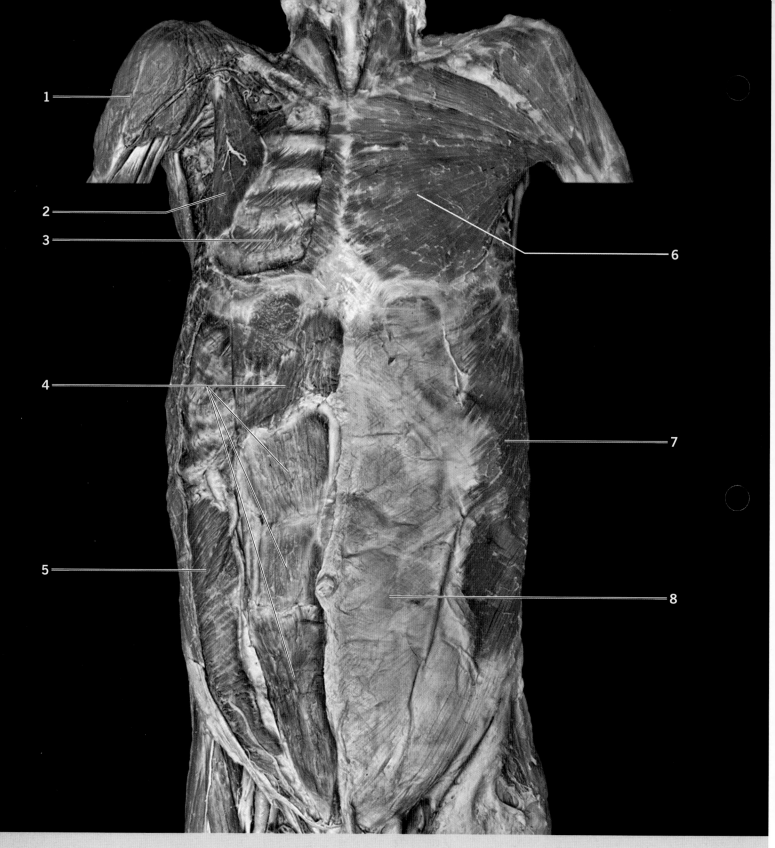

FIGURE 4.9 Intermediate Muscles of the Trunk, Anterior View

1. Deltoid
2. Pectoralis minor
3. Internal intercostal

4. Rectus abdominis
5. Internal oblique
6. Pectoralis major

7. External oblique
8. *Rectus sheath, anterior layer*

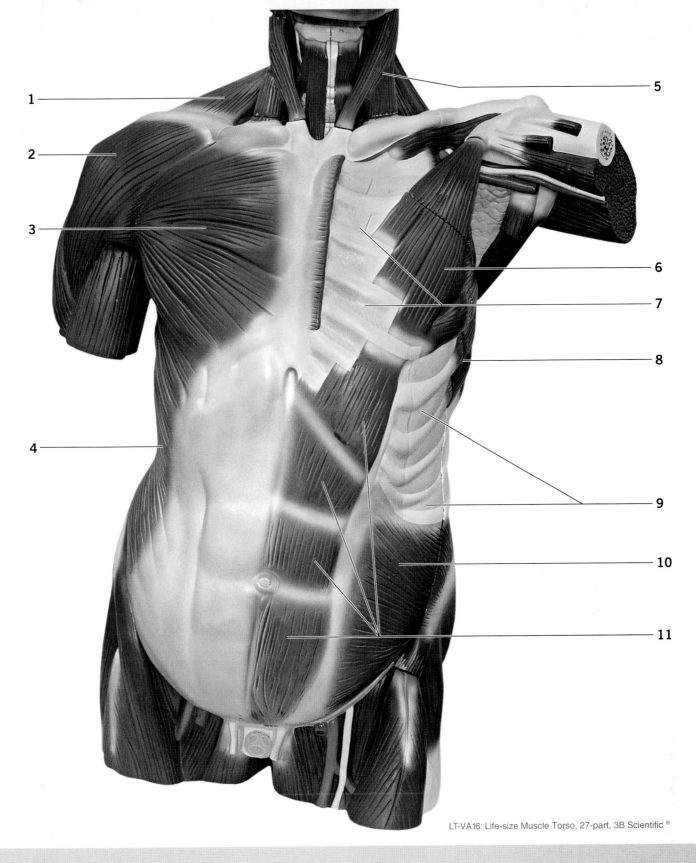

1
2
3
4
5
6
7
8
9
10
11

LT-VA16: Life-size Muscle Torso, 27-part, 3B Scientific ®

FIGURE 4.10 Superficial and Intermediate Muscles of the Trunk, Anterior View

1. Trapezius
2. Deltoid
3. Pectoralis major
4. External oblique
5. Sternocleidomastoid
6. Pectoralis minor
7. Internal intercostals
8. Serratus anterior
9. External intercostals
10. Internal oblique
11. Rectus abdominis

Muscles of the Trunk

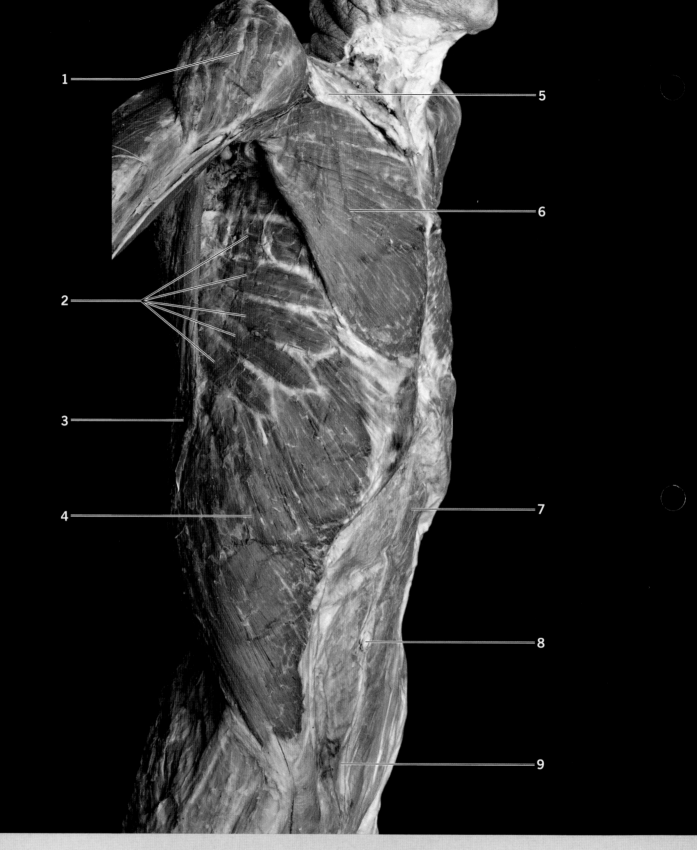

FIGURE 4.11 Superficial Muscles of the Trunk, Lateral View

1. Deltoid
2. Serratus anterior
3. Latissimus dorsi
4. External oblique
5. *Clavicle*
6. Pectoralis major
7. *Rectus sheath, anterior layer*
8. *Umbilicus*
9. *Linea alba*

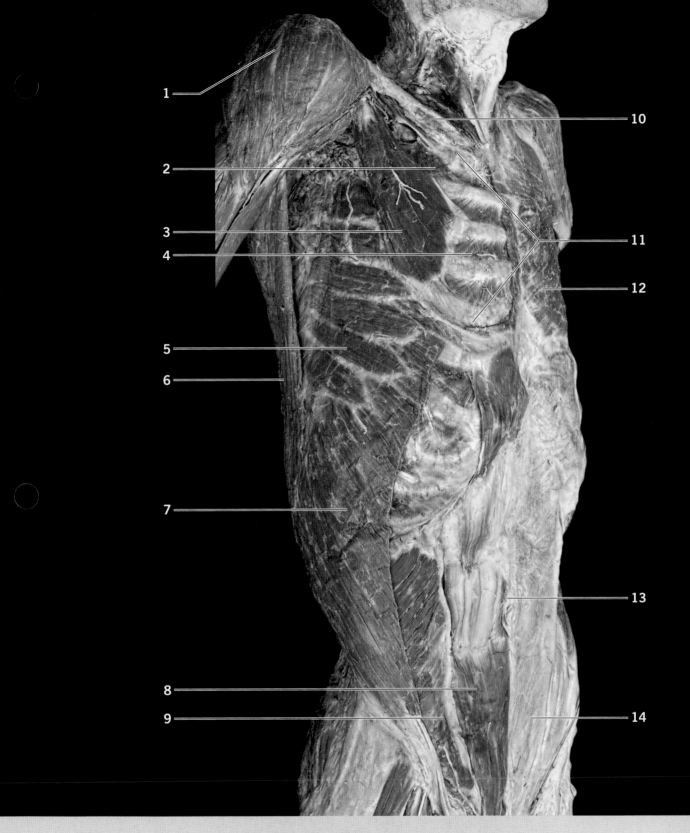

FIGURE 4.12 Intermediate Muscles of the Trunk, Lateral View

1. Deltoid
2. External intercostal
3. Pectoralis minor
4. Internal intercostal
5. Serratus anterior
6. Latissimus dorsi
7. External oblique
8. Rectus abdominis
9. Internal oblique
10. *Clavicle*
11. Costal cartilages
12. Pectoralis major
13. *Linea alba*
14. *Rectus sheath, anterior layer*

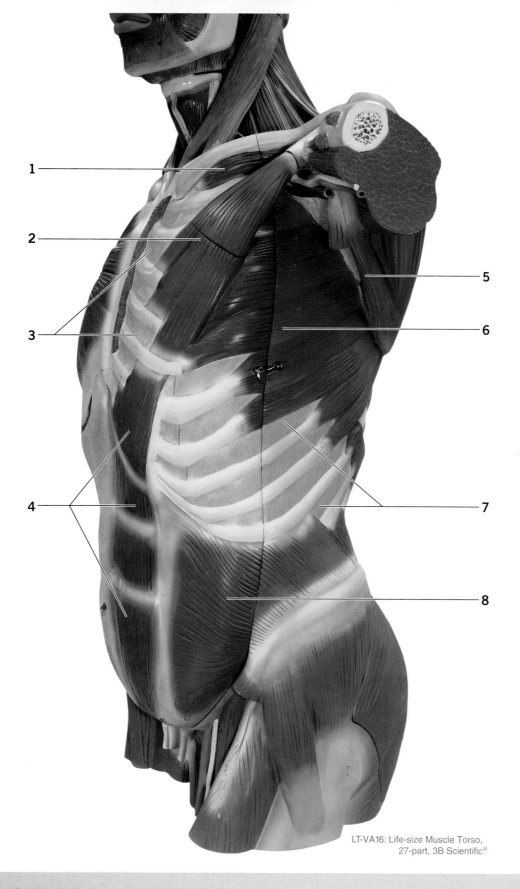

LT-VA16: Life-size Muscle Torso,
27-part, 3B Scientific®

FIGURE 4.13 Superficial and Intermediate Muscles of the Trunk, Lateral View

1. Subclavius
2. Pectoralis minor
3. Internal intercostals
4. Rectus abdominis
5. Teres major
6. Serratus anterior
7. External intercostals
8. Internal oblique

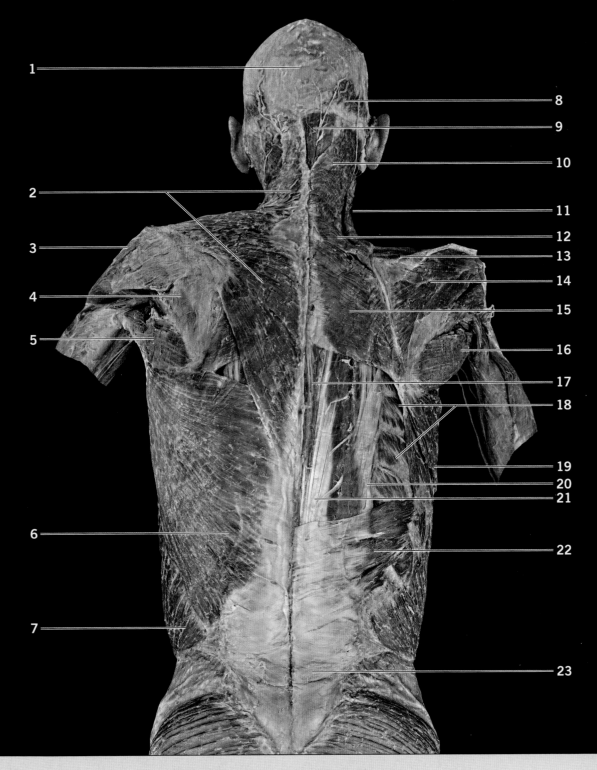

FIGURE 4.14 Superficial (Left) and Intermediate (Right) Muscles of the Trunk, Posterior View

1. Epicranial aponeurosis
2. Trapezius
3. Deltoid
4. Infraspinatus fascia
5. Teres major
6. Latissimus dorsi
7. External oblique
8. Epicranius, occipital belly
9. Semispinalis capitis
10. Splenius capitis
11. Levator scapulae
12. Rhomboid minor
13. Supraspinatus
14. Infraspinatus
15. Rhomboid major
16. Teres major
17. Spinalis
18. External intercostals
19. Serratus anterior
20. Iliocostalis
21. Longissimus
22. Serratus posterior inferior
23. Thoracolumbar fascia

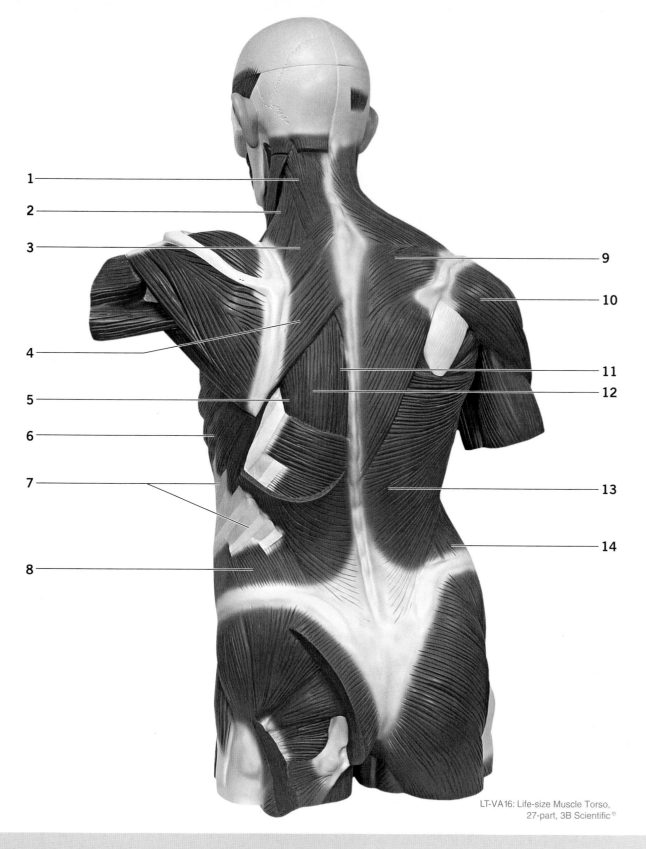

FIGURE 4.15 Superficial and Intermediate Muscles of the Trunk, Posterior View

LT-VA16: Life-size Muscle Torso,
27-part, 3B Scientific®

1. Splenius capitis
2. Levator scapulae
3. Rhomboid minor
4. Rhomboid major

5. Iliocostalis
6. Serratus anterior
7. External intercostals
8. Internal oblique

9. Trapezius
10. Deltoid
11. Spinalis
12. Longissimus

13. Latissimus dorsi
14. External oblique

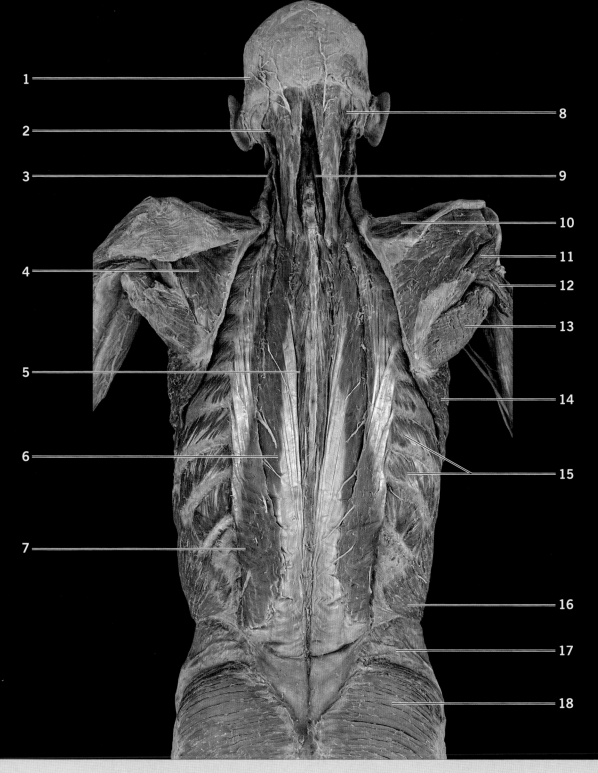

FIGURE 4.16 Intermediate Muscles of the Trunk, Posterior View

1. Epicranius, occipital belly
2. Semispinalis capitis
3. Levator scapulae
4. Infraspinatus
5. Spinalis
6. Longissimus
7. Iliocostalis
8. Obliquus capitis superior
9. Semispinalis cervicis
10. Supraspinatus
11. Teres minor
12. Triceps brachii, long head
13. Teres major
14. Serratus anterior
15. External intercostals
16. External oblique
17. Gluteus medius
18. Gluteus maximus

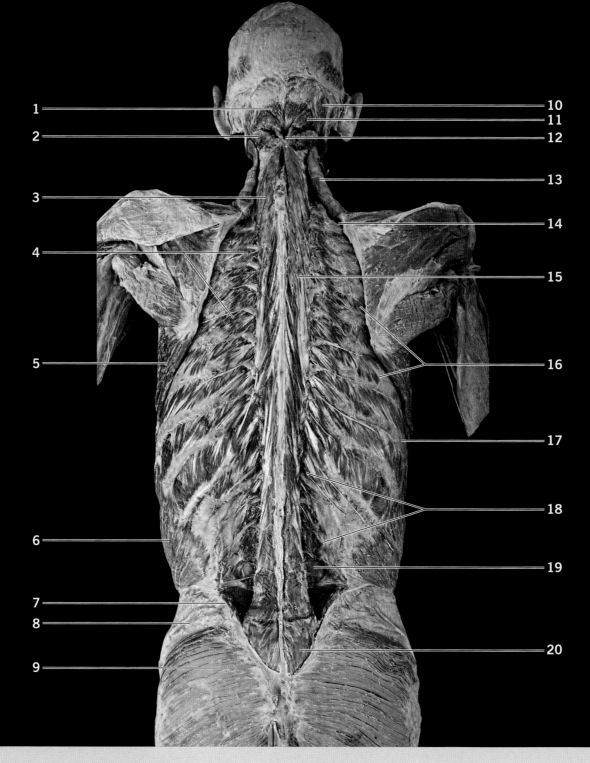

FIGURE 4.17 Deep Muscles of the Trunk, Posterior View

1. Rectus capitis posterior minor
2. Obliquus capitis inferior
3. Semispinalis cervicis
4. Levatores costarum
5. Serratus anterior
6. External oblique
7. *Iliac crest*
8. Gluteus medius
9. Gluteus maximus
10. Obliquus capitis superior
11. Rectus capitis posterior major
12. *Spinous process of axis*
13. Levator scapulae
14. *Rib 2*
15. Semispinalis thoracis
16. External intercostals
17. *Rib 9*
18. Intertransversarii
19. *Transverse process of L4*
20. Multifidus

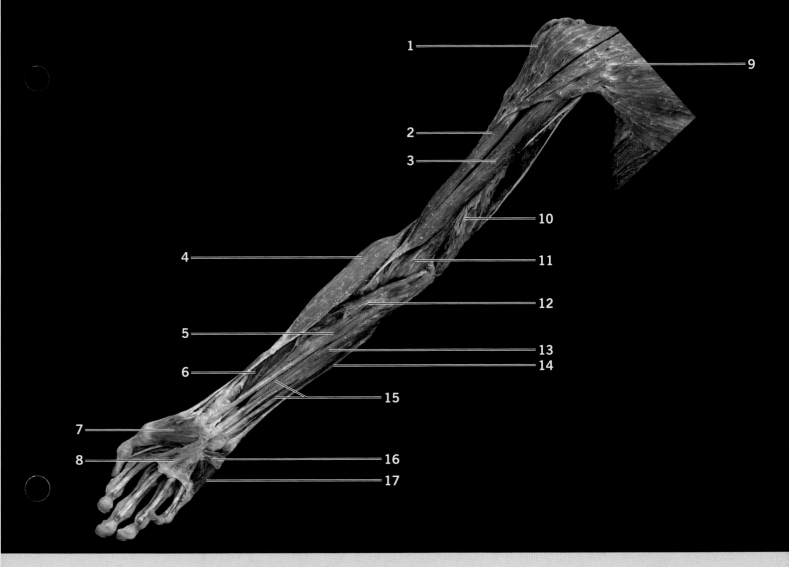

FIGURE 4.18 Superficial Muscles of Upper Limb, Anterior View, Right Side

1. Deltoid
2. Biceps brachii, long head
3. Biceps brachii, short head
4. Brachioradialis
5. Flexor carpi radialis
6. Flexor pollicis longus
7. Abductor pollicis brevis
8. *Palmar aponeurosis*
9. Pectoralis major
10. Triceps brachii, medial head
11. Brachialis
12. Pronator teres
13. Palmaris longus
14. Flexor carpi ulnaris
15. Flexor digitorum superficialis
16. Palmaris brevis
17. Abductor digiti minimi

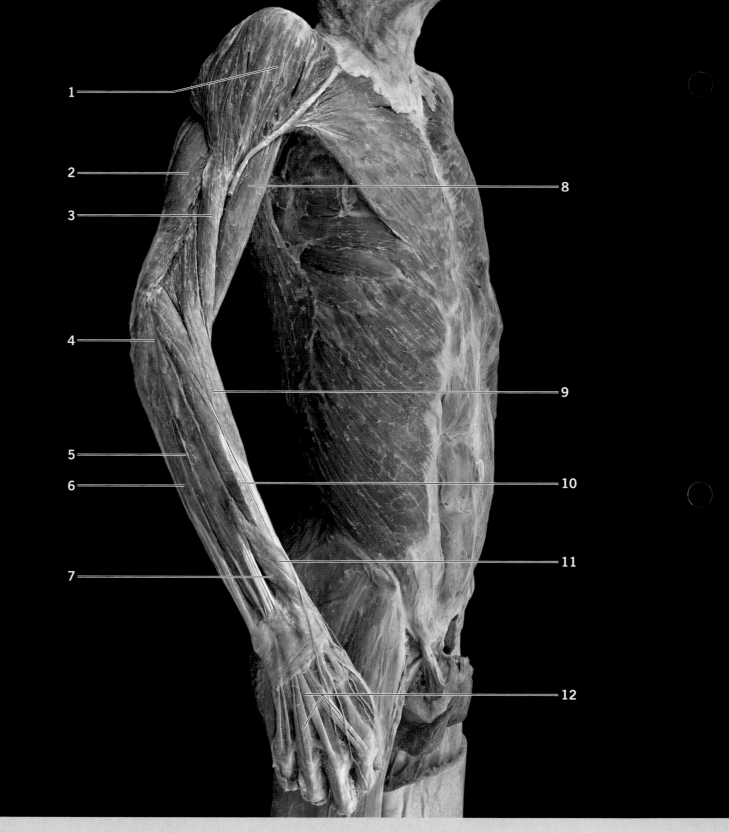

FIGURE 4.19 Muscles of the Upper Limb, Lateral View, Right Side

1. Deltoid
2. Triceps brachii, lateral head
3. Brachialis
4. Anconeus
5. Extensor digiti minimi
6. Extensor carpi ulnaris
7. Extensor pollicis brevis
8. Biceps brachii
9. Brachioradialis
10. Extensor carpi radialis brevis
11. Abductor pollicis longus
12. Extensor digitorum

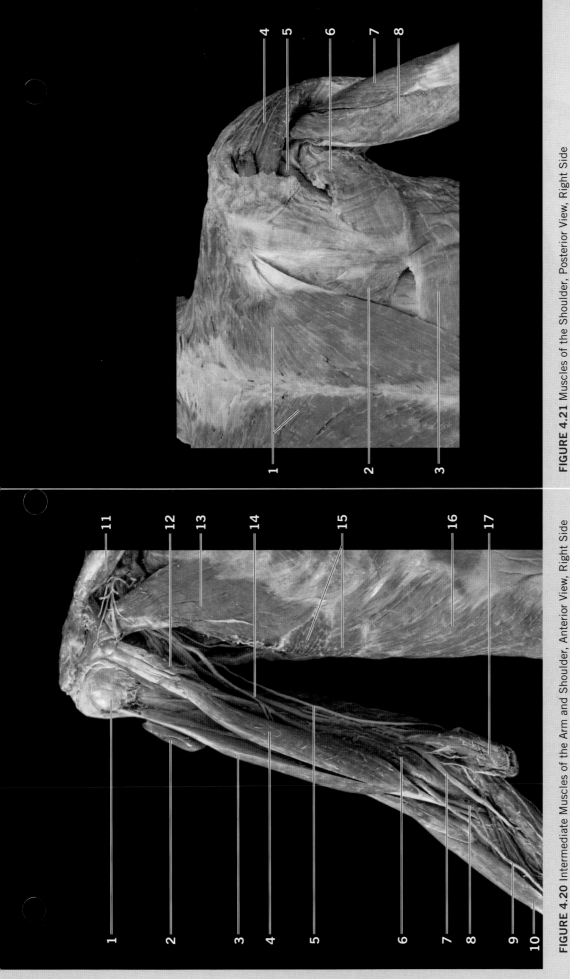

FIGURE 4.20 Intermediate Muscles of the Arm and Shoulder, Anterior View, Right Side

1. *Head of humerus within joint capsule*
2. Pectoralis major (cut)
3. Biceps brachii, long head
4. Biceps brachii, short head
5. *Ulnar nerve*
6. Brachialis
7. *Median nerve*
8. *Ulnar artery*
9. *Radial artery*
10. Brachioradialis
11. Subclavius
12. Coracobrachialis
13. Pectoralis minor
14. *Brachial artery*
15. Serratus anterior
16. External oblique
17. Pronator teres and flexor carpi radialis, reflected

FIGURE 4.21 Muscles of the Shoulder, Posterior View, Right Side

1. Trapezius
2. Rhomboid major
3. Latissimus dorsi
4. Deltoid
5. Teres minor
6. Teres major
7. Triceps brachii, lateral head
8. Triceps brachii, long head

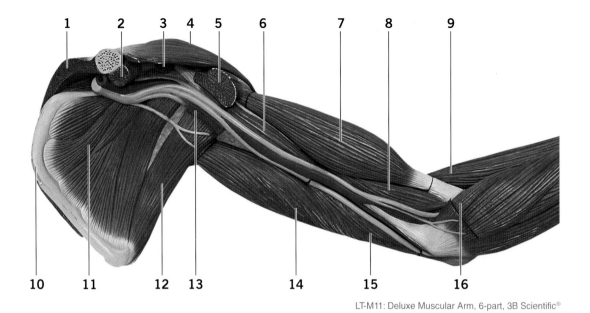

LT-M11: Deluxe Muscular Arm, 6-part, 3B Scientific®

FIGURE 4.22 Muscles of the Shoulder and Arm, Anterior View, Left Side

1. Trapezius (cut)
2. Subclavius (cut)
3. Pectoralis minor (cut)
4. Deltoid
5. Pectoralis major (cut)
6. Coracobrachialis
7. Biceps brachii
8. Brachialis
9. Brachioradialis
10. *Scapula*
11. Subscapularis
12. Teres major
13. Latissimus dorsi (cut)
14. Triceps brachii, long head
15. Triceps brachii, medial head
16. Pronator teres

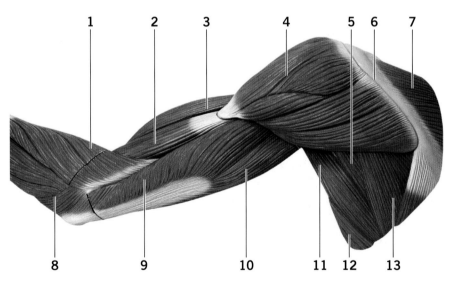

LT-M11: Deluxe Muscular Arm, 6-part, 3B Scientific®

FIGURE 4.23 Muscles of the Shoulder and Arm, Posterior View, Left Side

1. Brachioradialis
2. Brachialis
3. Biceps brachii
4. Deltoid
5. Teres minor
6. *Spine of scapula*
7. Trapezius
8. Anconeus
9. Triceps brachii, lateral head
10. Triceps brachii, long head
11. Teres major
12. Latissimus dorsi
13. Infraspinatus

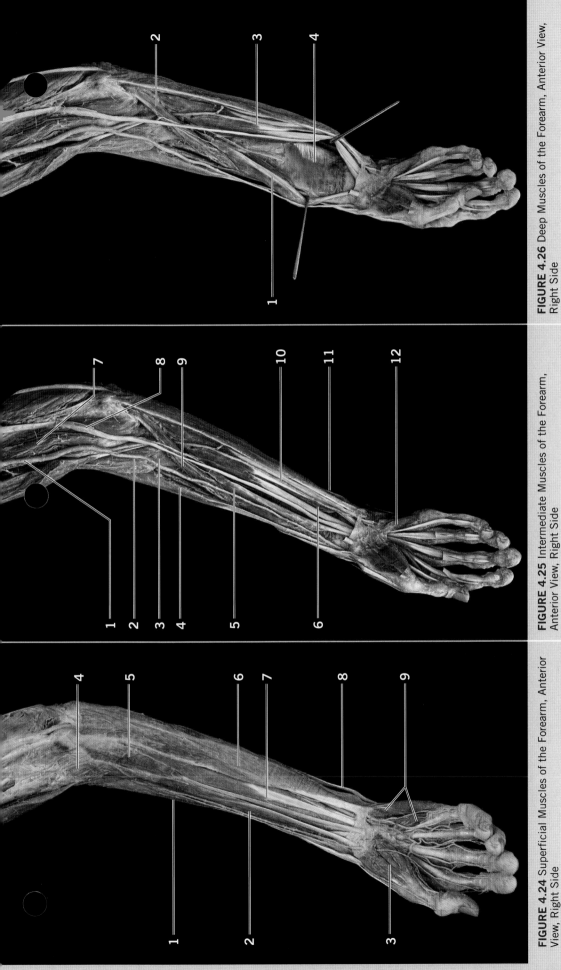

FIGURE 4.24 Superficial Muscles of the Forearm, Anterior View, Right Side

1. Brachioradialis
2. Flexor carpi radialis
3. Abductor pollicis brevis
4. *Bicipital aponeurosis*
5. Palmaris longus
6. Flexor carpi ulnaris
7. Flexor digitorum superficialis
8. Extensor carpi ulnaris
9. Hypothenar muscles

FIGURE 4.25 Intermediate Muscles of the Forearm, Anterior View, Right Side

1. *Brachial artery*
2. *Radial artery*
3. Supinator
4. Brachioradialis
5. Flexor pollicis longus
6. Flexor digitorum profundus
7. Brachialis
8. *Median nerve*
9. Pronator teres
10. *Ulnar artery*
11. Flexor carpi ulnaris
12. Flexor digiti minimi brevis

FIGURE 4.26 Deep Muscles of the Forearm, Anterior View, Right Side

1. Flexor pollicis longus
2. Pronator teres
3. Flexor digitorum profundus
4. Pronator quadratus

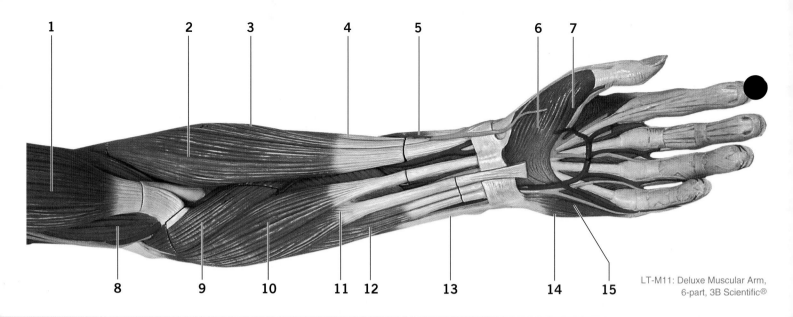

FIGURE 4.27 Superficial Muscles of the Forearm and Hand, Anterior View, Left Side

LT-M11: Deluxe Muscular Arm, 6-part, 3B Scientific®

1. Biceps brachii
2. Brachioradialis
3. Extensor carpi radialis longus
4. Extensor carpi radialis brevis
5. Abductor pollicis longus
6. Abductor pollicis brevis
7. Flexor pollicis brevis
8. Brachialis
9. Pronator teres
10. Flexor carpi radialis
11. Palmaris longus
12. Flexor digitorum superficialis
13. Flexor carpi ulnaris
14. Abductor digiti minimi
15. Flexor digiti minimi brevis

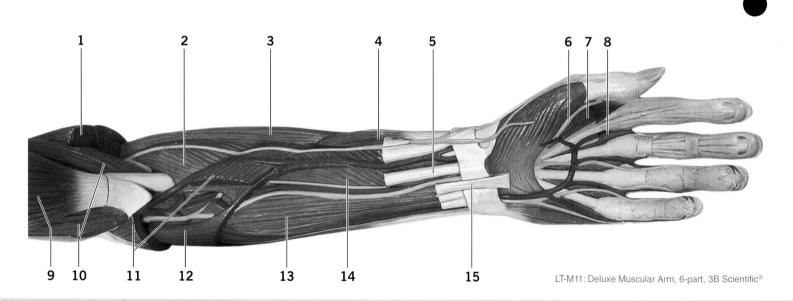

LT-M11: Deluxe Muscular Arm, 6-part, 3B Scientific®

FIGURE 4.28 Deep Muscles of the Forearm and Hand, Anterior View, Left Side

1. Brachioradialis (cut)
2. Supinator
3. Extensor digitorum
4. Abductor pollicis longus
5. Flexor carpi radialis tendon
6. Flexor pollicis longus tendon
7. Adductor pollicis
8. Lumbrical
9. Biceps brachii
10. Brachialis
11. Pronator teres (cut)
12. Flexor digitorum superficialis
13. Flexor digitorum profundus
14. Flexor pollicis longus
15. Palmaris longus tendon

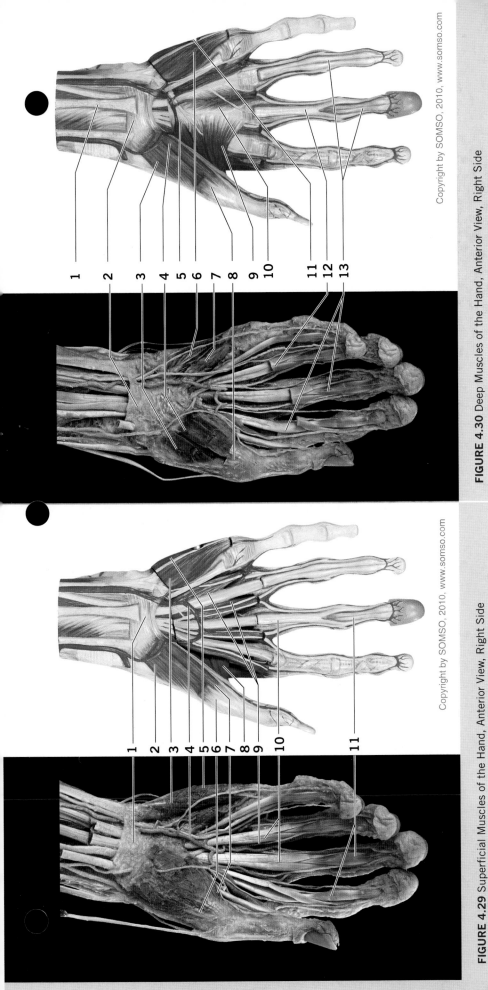

Copyright by SOMSO, 2010, www.somso.com

FIGURE 4.30 Deep Muscles of the Hand, Anterior View, Right Side

1. Tendon of palmaris longus
2. Flexor retinaculum
3. Opponens pollicis
4. Flexor pollicis brevis
5. *Deep palmar arch*
6. Flexor digiti minimi brevis
7. Opponens digiti minimi
8. Abductor pollicis brevis (cut)

9. Adductor pollicis
10. Dorsal interosseous
11. Abductor digiti minimi
12. Tendons of flexor digitorum superficialis
13. Tendons of flexor digitorum profundus

Copyright by SOMSO, 2010, www.somso.com

FIGURE 4.29 Superficial Muscles of the Hand, Anterior View, Right Side

1. Flexor retinaculum
2. Opponens pollicis
3. Flexor digiti minimi brevis
4. *Superficial palmar arch*
5. Abductor digiti minimi
6. Abductor pollicis brevis
7. Flexor pollicis brevis

8. Adductor pollicis
9. Lumbricals
10. Tendons of flexor digitorum superficialis
11. Tendons of flexor digitorum profundus

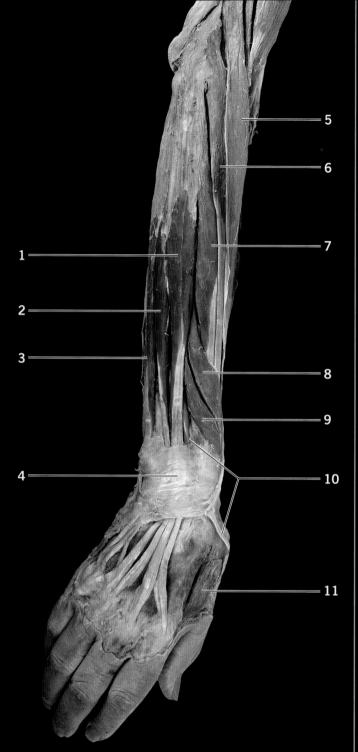

FIGURE 4.31 Superficial Muscles of the Forearm, Posterior View, Right Side

FIGURE 4.32 Intermediate Muscles of the Forearm, Posterior View, Right Side

1. Extensor digitorum
2. Extensor digiti minimi
3. Extensor carpi ulnaris
4. Extensor retinaculum
5. Brachioradialis
6. Extensor carpi radialis longus
7. Extensor carpi radialis brevis
8. Abductor pollicis longus
9. Extensor pollicis brevis
10. Extensor pollicis longus
11. 1st dorsal interosseous

1. Extensor digitorum (cut)
2. Extensor carpi ulnaris
3. Extensor digiti minimi
4. Brachioradialis
5. Extensor carpi radialis longus
6. Extensor carpi radialis brevis
7. Abductor pollicis longus
8. Extensor pollicis brevis
9. Extensor pollicis longus
10. Extensor indicis

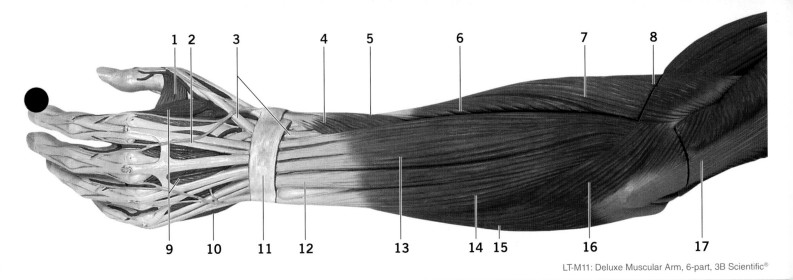

FIGURE 4.33 Superficial Muscles of Forearm and Hand, Posterior View, Left Side

1. Adductor pollicis
2. Extensor indicis tendon
3. Extensor pollicis longus
4. Extensor pollicis brevis
5. Abductor pollicis longus
6. Extensor carpi radialis brevis
7. Extensor carpi radialis longus
8. Brachioradialis
9. 1st dorsal interosseous
10. Abductor digiti minimi
11. Extensor retinaculum
12. Extensor digiti minimi
13. Extensor digitorum
14. Extensor carpi ulnaris
15. Flexor carpi ulnaris
16. Anconeus
17. Triceps brachii

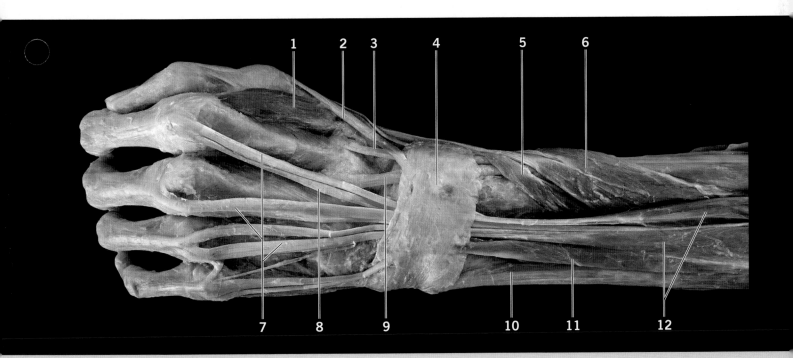

FIGURE 4.34 Muscles of the Hand, Posterior View, Left Side

1. 1st dorsal interosseous
2. Tendon of extensor pollicis longus
3. Tendon of extensor carpi radialis longus
4. Extensor retinaculum
5. Extensor pollicis brevis
6. Abductor pollicis longus
7. Tendons of extensor digitorum
8. Tendon of extensor indicis
9. Tendon of extensor carpi radialis brevis
10. Extensor carpi ulnaris
11. Extensor digiti minimi
12. Extensor digitorum

Muscles of the Upper Limb

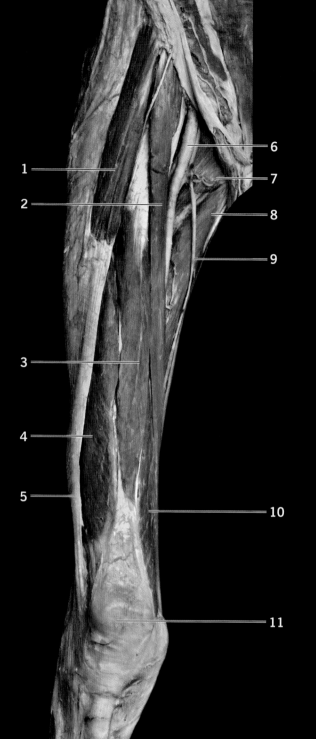

FIGURE 4.35 Superficial Muscles of the Thigh, Anterior View, Right Side

FIGURE 4.36 Intermediate Muscles of the Thigh, Anterior View, Right Side

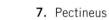
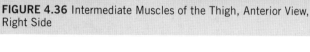
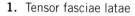

1. Tensor fasciae latae	7. Pectineus
2. Sartorius	8. Adductor longus
3. Rectus femoris	9. *Great saphenous vein*
4. Vastus lateralis	10. Vastus medialis
5. *Iliotibial tract*	11. *Patella*
6. *Femoral artery*	

1. Tensor fasciae latae	8. *Femoral nerve*
2. Iliopsoas	9. Pectineus
3. Sartorius (cut)	10. Adductor brevis
4. Vastus intermedius	11. *Femoral artery*
5. *Iliotibial tract*	12. Adductor longus (cut)
6. Vastus lateralis	13. Gracilis
7. Rectus femoris (cut)	14. Vastus medialis

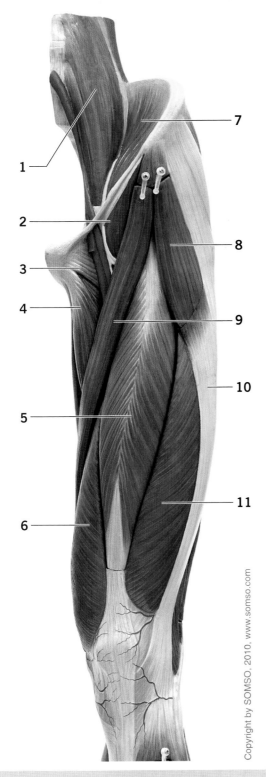

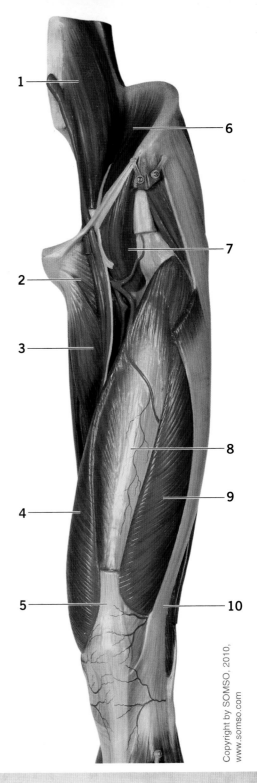

FIGURE 4.37 Superficial Muscles of the Thigh, Anterior View, Left Side

FIGURE 4.38 Deep Muscles of the Thigh, Anterior View, Left Side

Copyright by SOMSO, 2010, www.somso.com

Copyright by SOMSO, 2010, www.somso.com

1. Psoas major
2. Iliopsoas
3. Pectineus
4. Adductor longus
5. Rectus femoris
6. Vastus medialis
7. Iliacus
8. Tensor fasciae latae
9. Sartorius
10. *Iliotibial tract*
11. Vastus lateralis

1. Psoas major
2. Pectineus
3. Adductor longus
4. Vastus medialis
5. Quadriceps femoris tendon
6. Iliacus
7. Iliopsoas
8. Vastus intermedius
9. Vastus lateralis
10. *Iliotibial tract*

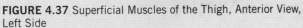

Muscles of the Lower Limb

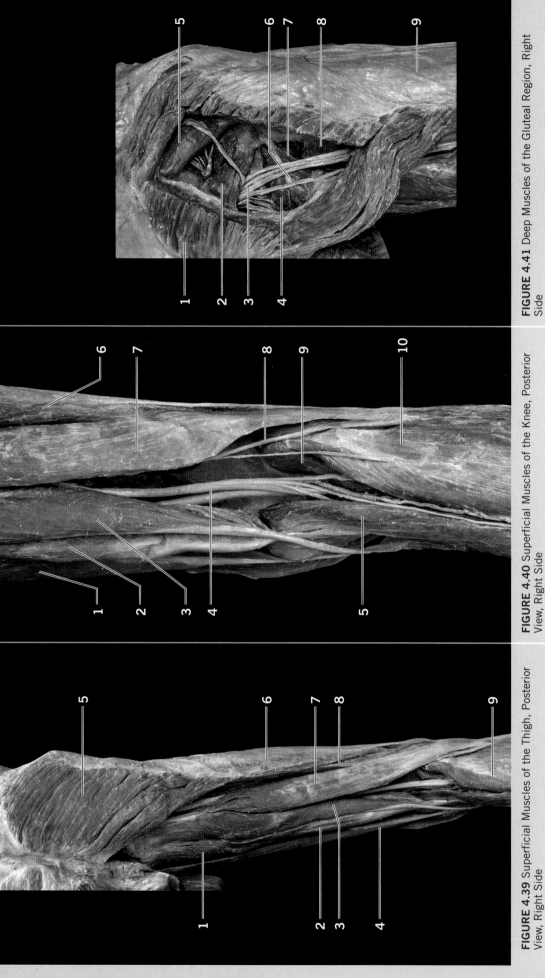

FIGURE 4.39 Superficial Muscles of the Thigh, Posterior View, Right Side

1. Adductor magnus
2. Semimembranosus
3. Semitendinosus
4. Gracilis
5. Gluteus maximus
6. Fascia lata
7. Biceps femoris, long head
8. Biceps femoris, short head
9. Gastrocnemius

FIGURE 4.40 Superficial Muscles of the Knee, Posterior View, Right Side

1. Gracilis
2. Semimembranosus
3. Semitendinosus
4. *Tibial nerve*
5. Gastrocnemius, medial head
6. Biceps femoris, short head
7. Biceps femoris, long head
8. *Common fibular nerve*
9. Plantaris muscle
10. Gastrocnemius, lateral head

FIGURE 4.41 Deep Muscles of the Gluteal Region, Right Side

1. Gluteus maximus
2. Piriformis
3. *Sciatic nerve*
4. Superior gemellus
5. Gluteus medius
6. Obturator internus
7. Inferior gemellus
8. Quadratus femoris
9. Iliotibial band

Muscles of the Lower Limb

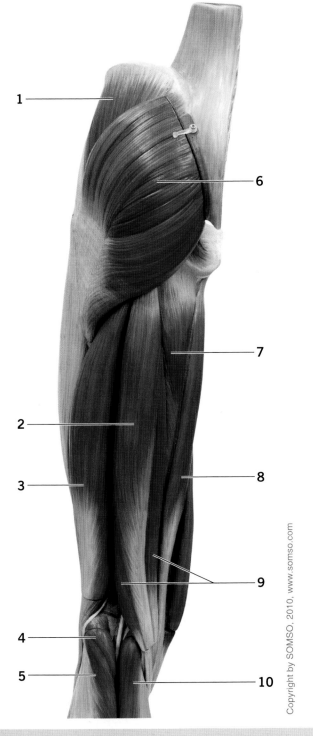

FIGURE 4.42 Superficial Muscles of the Thigh, Posterior View, Left Side

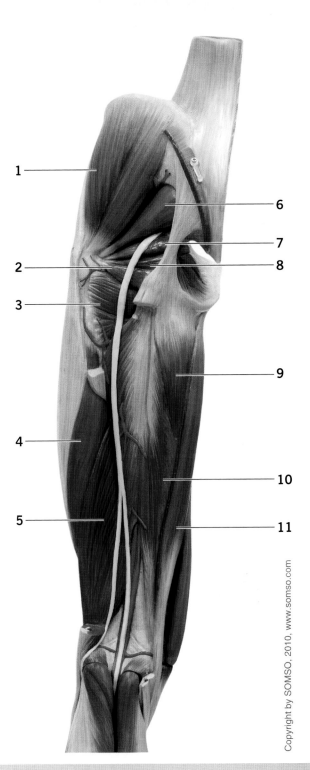

FIGURE 4.43 Deep Muscles of the Thigh, Posterior View, Left Side

Copyright by SOMSO, 2010, www.somso.com

1. Gluteus medius
2. Semitendinosus
3. Biceps femoris, long head
4. Plantaris
5. Gastrocnemius, lateral head
6. Gluteus maximus
7. Adductor magnus
8. Gracilis
9. Semimembranosus
10. Gastrocnemius, medial head

1. Gluteus medius
2. Gemellus inferior
3. Quadratus femoris
4. Vastus lateralis
5. Biceps femoris, short head
6. Piriformis
7. Gemellus superior
8. Obturator internus
9. Adductor magnus
10. Semimembranosus
11. Gracilis

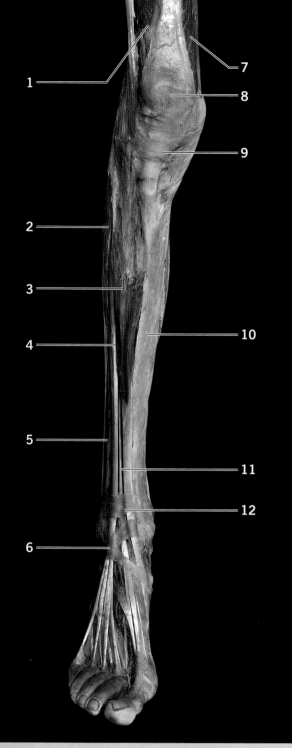

FIGURE 4.44 Superficial Muscles of the Leg, Anterior View, Right Side

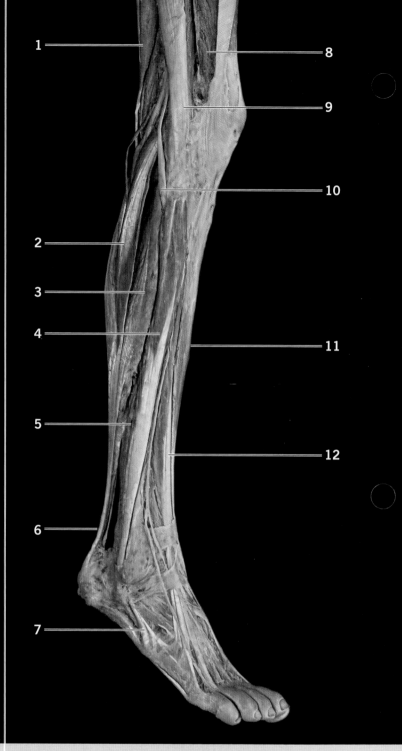

FIGURE 4.45 Superficial Muscles of the Leg, Lateral View, Side

1. Vastus lateralis
2. Fibularis longus
3. Tibialis anterior
4. Extensor digitorum longus
5. Fibularis brevis
6. Inferior extensor retinaculum
7. Vastus medialis
8. *Patella*
9. Patellar ligament
10. *Tibia*
11. Extensor hallucis longus
12. Superior extensor retinaculum

1. Semitendinosus
2. Gastrocnemius, lateral head
3. Soleus
4. Fibularis longus
5. Fibularis brevis
6. Calcaneal tendon
7. Fibularis brevis tendon
8. Vastus lateralis
9. Biceps femoris, long head
10. *Common fibular nerve*
11. Tibialis anterior
12. Extensor digitorum longus

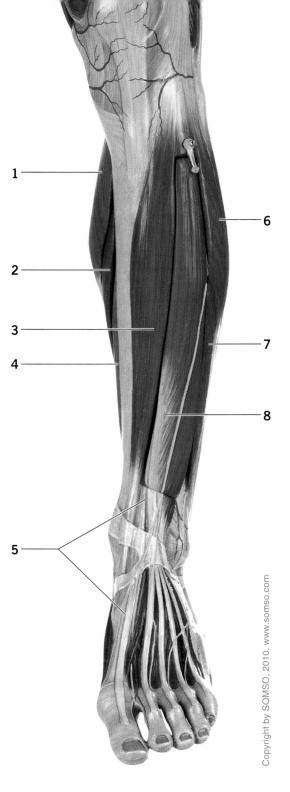

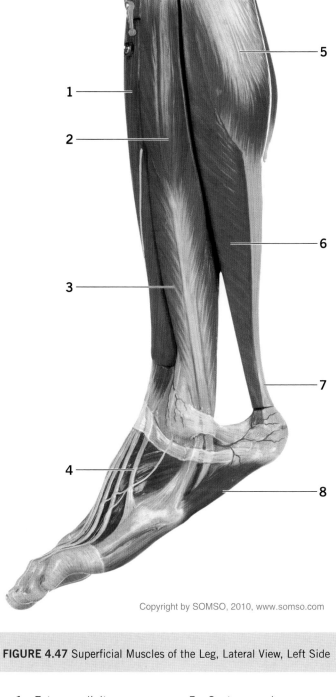

FIGURE 4.46 Superficial Muscles of the Leg, Anterior View, Left Side

Copyright by SOMSO, 2010, www.somso.com

FIGURE 4.47 Superficial Muscles of the Leg, Lateral View, Left Side

Copyright by SOMSO, 2010, www.somso.com

1. Gastrocnemius
2. Soleus
3. Tibialis anterior
4. Flexor digitorum longus
5. Extensor hallucis longus
6. Fibularis longus
7. Fibularis brevis
8. Extensor digitorum longus

1. Extensor digitorum longus
2. Fibularis longus
3. Fibularis brevis
4. Fibularis tertius
5. Gastrocnemius
6. Soleus
7. Calcaneal tendon
8. Abductor digiti minimi

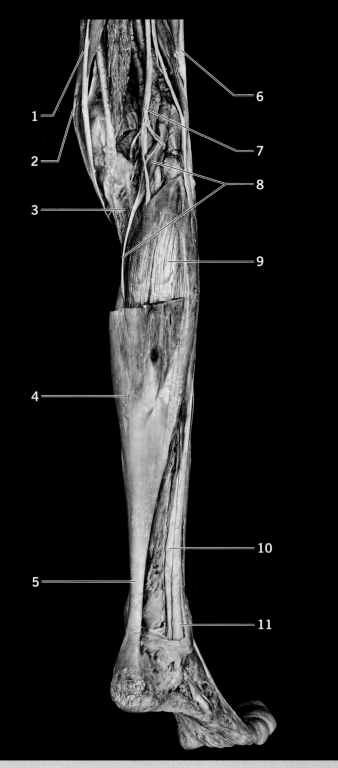

FIGURE 4.48 Intermediate Muscles of the Leg, Posterior View, Right Side

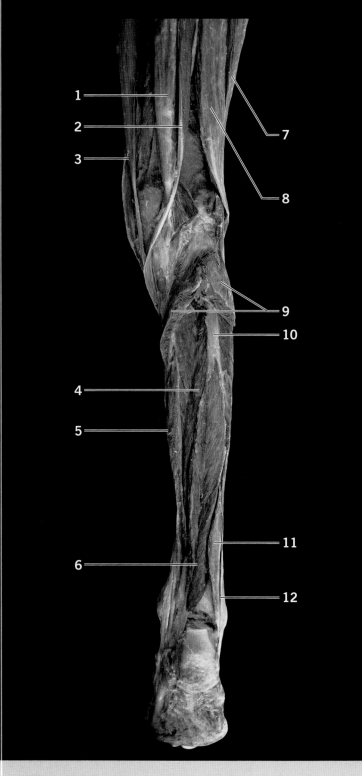

FIGURE 4.49 Deep Muscles of the Leg, Posterior View, Right Side

1. Semitendinosus
2. Gracilis
3. Popliteus
4. Gastrocnemius
5. Calcaneal tendon
6. Biceps femoris
7. *Tibial nerve*
8. Plantaris
9. Soleus
10. Fibularis brevis
11. Fibularis longus tendon

1. Semimembranosus
2. Semitendinosus
3. Gracilis
4. Tibialis posterior
5. Flexor digitorum longus
6. Flexor hallucis longus
7. Biceps femoris, short head
8. Biceps femoris, long head
9. Gastrocnemius (cut)
10. *Fibula*
11. Fibularis brevis
12. Fibularis longus

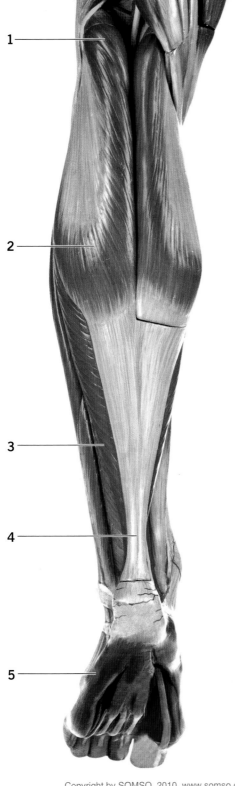

FIGURE 4.50 Superficial Muscles of the Leg, Posterior View, Left Side

FIGURE 4.51 Deep Muscles of the Leg, Posterior View, Left Side

1. Plantaris
2. Gastrocnemius
3. Soleus
4. Calcaneal tendon
5. Abductor digiti minimi

1. Fibularis longus
2. Flexor hallucis longus
3. Fibularis brevis
4. Abductor digiti minimi
5. Popliteus
6. Tibialis posterior
7. Flexor digitorum longus
8. Abductor hallucis

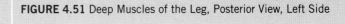

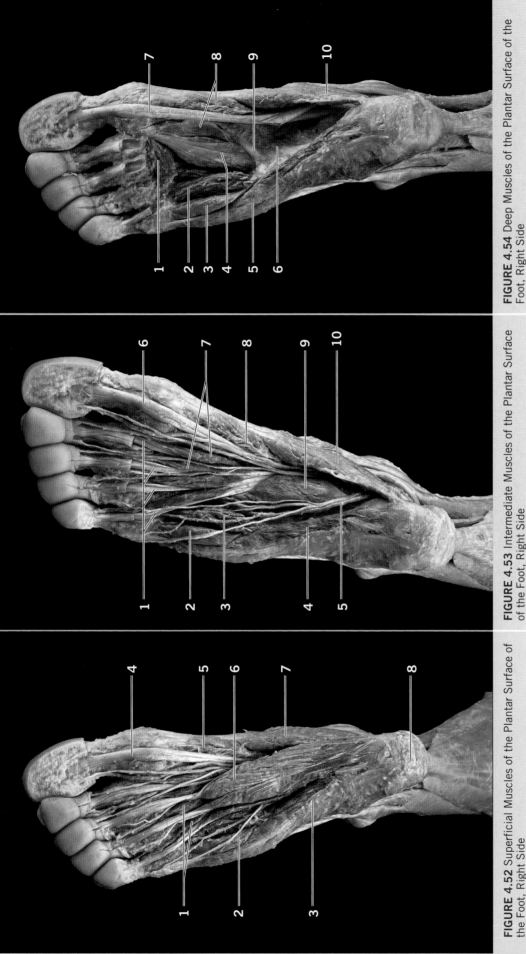

FIGURE 4.52 Superficial Muscles of the Plantar Surface of the Foot, Right Side

1. Lumbricals
2. Flexor digiti minimi brevis
3. Abductor digiti minimi
4. Flexor hallucis longus tendon
5. Flexor hallucis brevis
6. Flexor digitorum brevis
7. Abductor hallucis
8. *Calcaneal tuberosity*

FIGURE 4.53 Intermediate Muscles of the Plantar Surface of the Foot, Right Side

1. Flexor digitorum longus tendons
2. Flexor digiti minimi brevis
3. Plantar interosseus
4. Abductor digiti minimi
5. *Lateral plantar artery*
6. Flexor hallucis longus tendon
7. Lumbricals
8. Flexor hallucis brevis
9. Quadratus plantae
10. Abductor hallucis

FIGURE 4.54 Deep Muscles of the Plantar Surface of the Foot, Right Side

1. Adductor hallucis, transverse head
2. Plantar interosseus
3. Flexor digiti minimi brevis
4. Adductor hallucis, oblique head
5. Abductor digiti minimi
6. Long plantar ligament
7. Flexor hallucis longus tendon
8. Flexor hallucis brevis
9. Fibularis longus tendon
10. Abductor hallucis

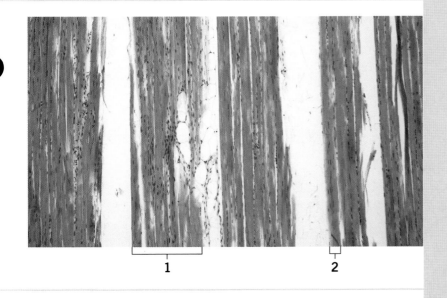

FIGURE 4.55 Skeletal Muscle Tissue, Longitudinal Section, 100×

1. Fascicle
2. Skeletal muscle fiber

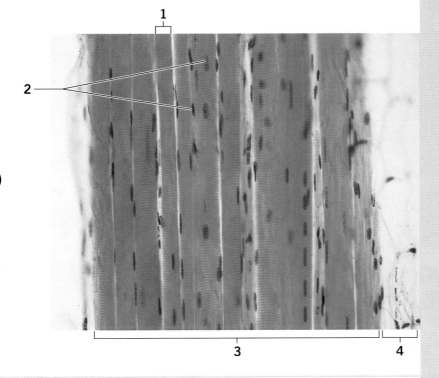

FIGURE 4.56 Skeletal Muscle Tissue, Longitudinal Section, 400×

1. Skeletal muscle fiber
2. Nuclei
3. Fascicle
4. Perimysium

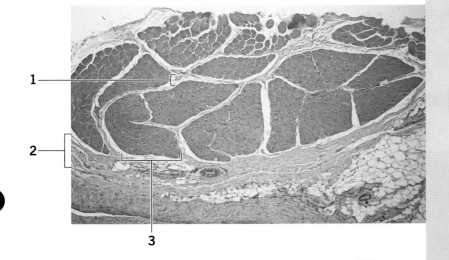

FIGURE 4.57 Skeletal Muscle Tissue, Cross Section, 100×

1. Perimysium
2. Epimysium
3. Fascicle

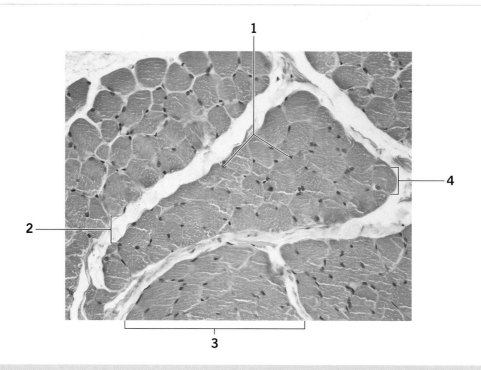

FIGURE 4.58 Skeletal Muscle Tissue, Cross Section, 400×

1. Nuclei   **2.** Perimysium   **3.** Fascicle   **4.** Skeletal muscle fiber

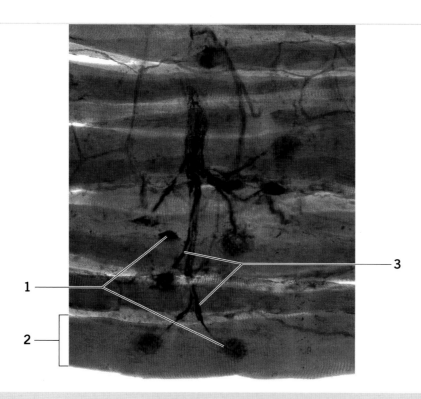

FIGURE 4.59 Neuromuscular Junction, 400×

1. Motor end plates   **2.** Skeletal muscle fiber   **3.** Motor axon terminals

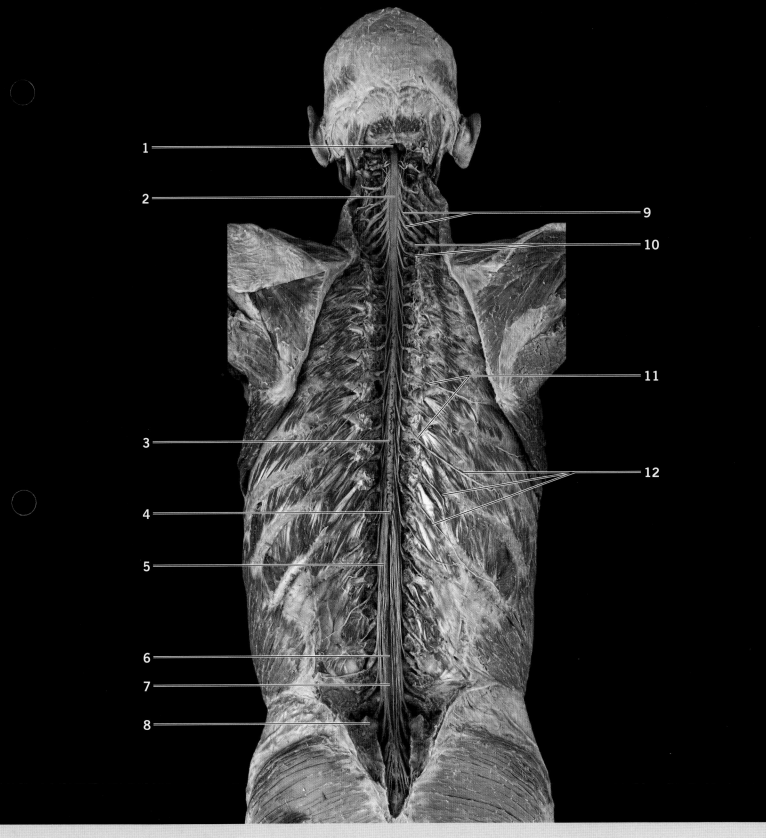

FIGURE 5.1 Spinal Cord

1. *Foramen magnum*
2. Cervical enlargement
3. Lumbar enlargement
4. Conus medullaris
5. Cauda equina
6. Filum terminale
7. Subarachnoid space
8. Superior articular process of sacrum
9. Dorsal roots
10. Dorsal root ganglia
11. Spinal nerves
12. Dorsal rami

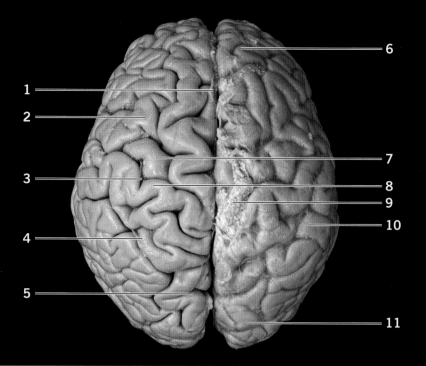

FIGURE 5.2 Brain, Superior View

1. Longitudinal fissure
2. Pia mater covering left cerebral hemisphere
3. Central sulcus
4. Parietal lobe
5. Parieto-occipital sulcus
6. Frontal lobe
7. Precentral gyrus
8. Postcentral gyrus
9. Arachnoid villi
10. Arachnoid mater covering right cerebral hemisphere
11. Occipital lobe

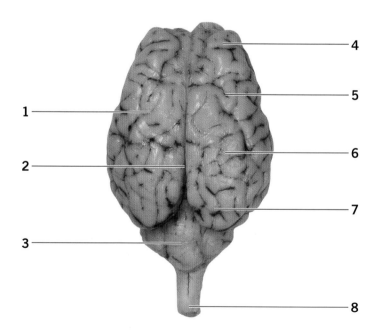

FIGURE 5.3 Sheep Brain, Dorsal View

1. Left cerebral hemisphere
2. Longitudinal fissure
3. Cerebellum
4. Frontal lobe
5. Right cerebral hemisphere
6. Parietal lobe
7. Occipital lobe
8. Spinal cord

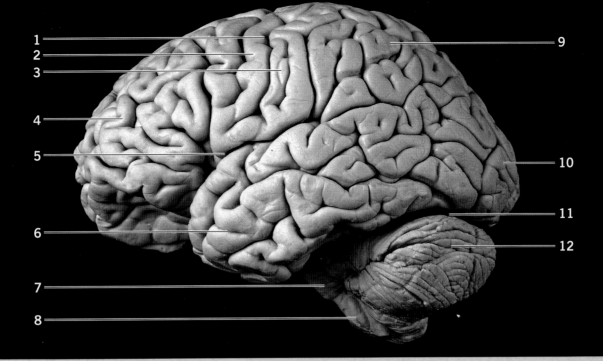

FIGURE 5.4 Brain, Lateral View

1. Central sulcus	**4.** Frontal lobe	**7.** Pons	**10.** Occipital lobe
2. Precentral gyrus	**5.** Lateral sulcus	**8.** Medulla oblongata	**11.** Transverse fissure
3. Postcentral gyrus	**6.** Temporal lobe	**9.** Parietal lobe	**12.** Cerebellum

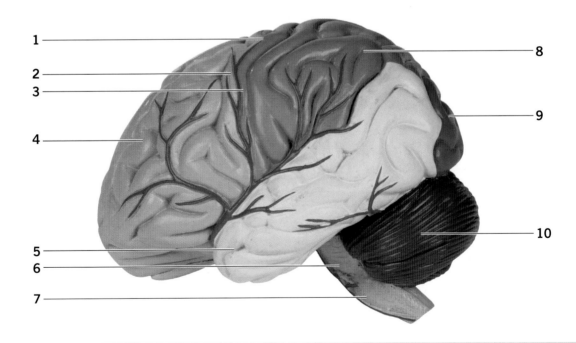

FIGURE 5.5 Brain, Lateral View

1. Central sulcus	**4.** Frontal lobe	**7.** Medulla oblongata	**10.** Cerebellum
2. Precentral gyrus	**5.** Temporal lobe	**8.** Parietal lobe	
3. Postcentral gyrus	**6.** Pons	**9.** Occipital lobe	

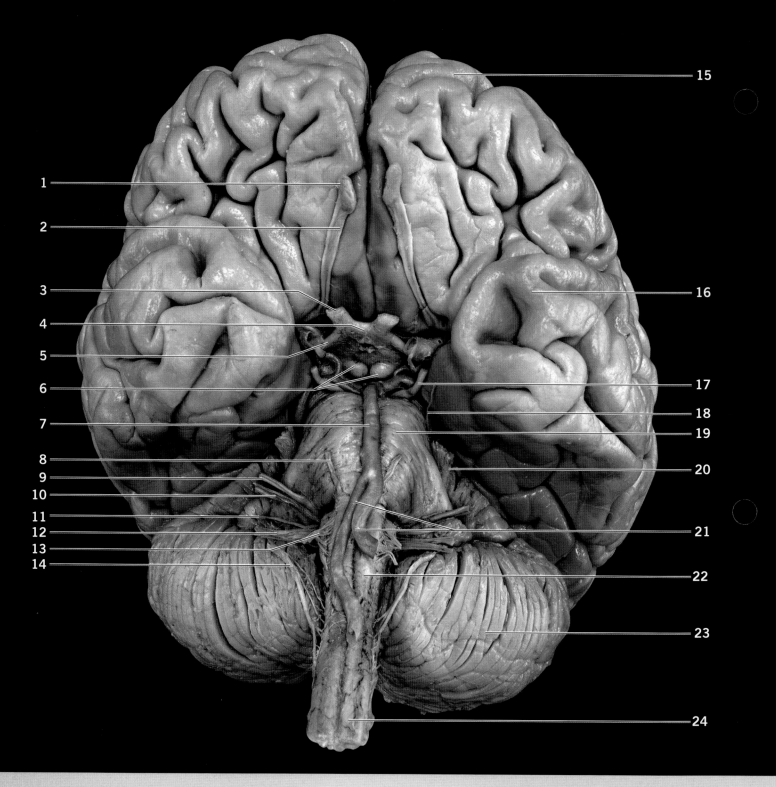

FIGURE 5.6 Brain with Cranial Nerves, Inferior View

1. Olfactory bulb
2. Olfactory tract
3. Optic nerve (II)
4. Optic chiasma
5. Optic tract
6. Mammillary bodies
7. *Basilar artery*

8. Abducens nerve (VI)
9. Facial nerve (VII)
10. Vestibulocochlear nerve (VIII)
11. Glossopharyngeal nerve (IX)
12. Vagus nerve (X)

13. Hypoglossal nerve (XII)
14. Accessory nerve (XI)
15. Frontal lobe
16. Temporal lobe
17. Oculomotor nerve (III)
18. Trochlear nerve (IV)
19. Pons

20. Trigeminal nerve (V)
21. *Vertebral arteries*
22. Medulla oblongata (pyramid)
23. Cerebellum
24. Spinal cord

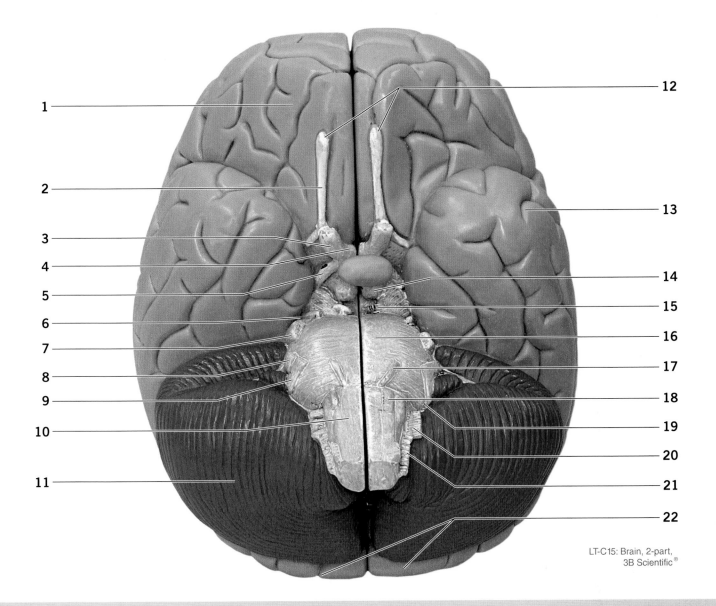

FIGURE 5.7 Brain with Cranial Nerves, Inferior View

LT-C15: Brain, 2-part,
3B Scientific®

1. Frontal lobe
2. Olfactory tract
3. Optic nerve (II)
4. Optic chiasma
5. Optic tract
6. Trochlear nerve (IV)

7. Trigeminal nerve (V)
8. Vestibulocochlear nerve (VIII)
9. Facial nerve (VII)
10. Medulla oblongata
11. Cerebellum

12. Olfactory bulbs
13. Temporal lobe
14. Mammillary body
15. Oculomotor nerve (III)
16. Pons
17. Abducens nerve (VI)

18. Hypoglossal nerve (XII)
19. Glossopharyngeal nerve (IX)
20. Vagus nerve (X)
21. Accessory nerve (XI)
22. Occipital lobes

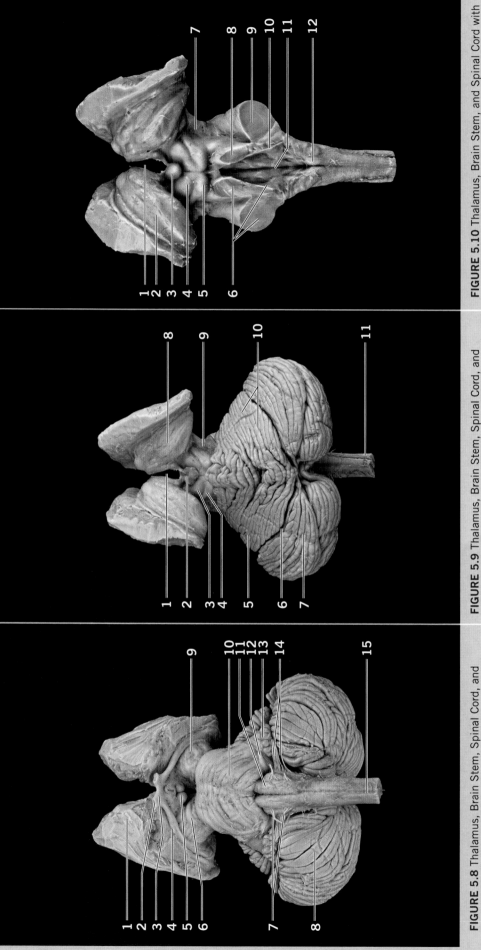

FIGURE 5.8 Thalamus, Brain Stem, Spinal Cord, and Cerebellum, Anterior View

1. Thalamus
2. Optic nerve (II)
3. Optic chiasma
4. Optic tract
5. Mammillary body
6. Oculomotor nerve (III)
7. Hypoglossal nerve (XII)
8. Cerebellum
9. Midbrain (cerebral peduncle)
10. Pons
11. Medulla oblongata
12. Vestibulocochlear nerve (VIII)
13. Facial nerve (VII)
14. Glossopharyngeal and vagus nerves (IX, X)
15. Spinal cord

FIGURE 5.9 Thalamus, Brain Stem, Spinal Cord, and Cerebellum, Posterior View

1. Third ventricle
2. Pineal gland
3. Superior colliculus
4. Inferior colliculus
5. Anterior lobe
6. Primary fissure
7. Posterior lobe
8. Thalamus
9. Cerebral peduncle
10. Folia
11. Spinal cord

FIGURE 5.10 Thalamus, Brain Stem, and Spinal Cord with Cerebellum Removed, Posterior View

1. Third ventricle
2. Thalamus
3. Pineal gland
4. Superior colliculus
5. Inferior colliculus
6. Cerebellar peduncles
7. Cerebral peduncle
8. Superior cerebellar peduncle
9. Middle cerebellar peduncle
10. Inferior cerebellar peduncle
11. Floor of fourth ventricle
12. Medulla oblongata

Central Nervous System

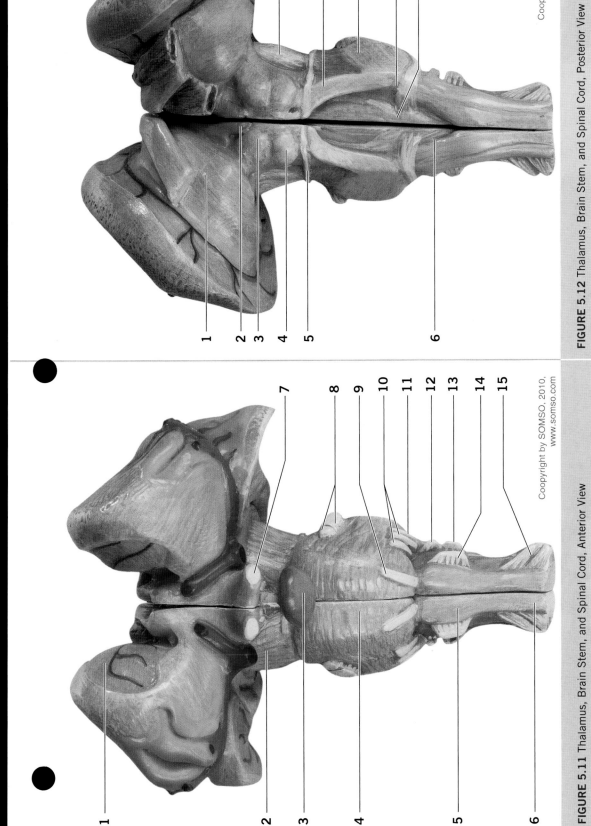

FIGURE 5.12 Thalamus, Brain Stem, and Spinal Cord, Posterior View

1. Thalamus
2. Pineal gland
3. Superior colliculus
4. Inferior colliculus
5. Trochlear nerve (IV)
6. Medulla oblongata

7. Cerebral peduncle
8. Superior cerebellar peduncle
9. Middle cerebellar peduncle
10. Inferior cerebellar peduncle
11. Floor of fourth ventricle

FIGURE 5.11 Thalamus, Brain Stem, and Spinal Cord, Anterior View

1. Thalamus
2. Midbrain (cerebral peduncle)
3. Pituitary gland
4. Pons
5. Medulla oblongata
6. Spinal cord
7. Optic nerve (II)
8. Trigeminal nerve (V)

9. Abducens nerve (VI)
10. Facial nerve (VII)
11. Vestibulocochlear nerve (VII)
12. Glossopharyngeal nerve (IX)
13. Vagus nerve (X)
14. Hypoglossal nerve (XII)
15. Accessory nerve (XI)

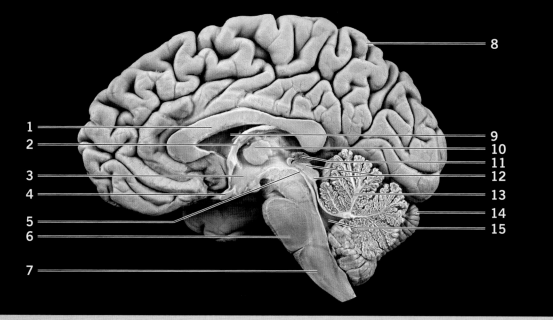

FIGURE 5.13 Brain, Midsagittal View

1. Corpus callosum
2. Fornix
3. Hypothalamus
4. Optic chiasma
5. Cerebral aqueduct
6. Pons
7. Medulla oblongata
8. Cerebrum
9. Lateral ventricle
10. Interthalamic adhesion
11. Pineal gland
12. Corpora quadrigemina
13. Arbor vitae (of cerebellum)
14. Cerebellum
15. Fourth ventricle

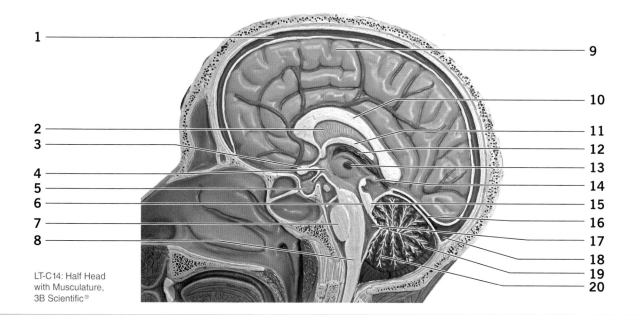

LT-C14: Half Head
with Musculature,
3B Scientific®

FIGURE 5.14 Brain, Midsagittal View

1. Superior sagittal sinus
2. Septum pellucidum
3. Optic chiasma
4. Hypothalamus
5. Pituitary gland
6. Cerebral aqueduct
7. Pons
8. Medulla oblongata
9. Cerebrum
10. Corpus callosum
11. Fornix
12. Choroid plexus of third ventricle
13. Interthalamic adhesion
14. Pineal gland
15. Corpora quadrigemina
16. Straight sinus
17. Fourth ventricle
18. Confluence of the sinuses
19. Cerebellum
20. Arbor vitae (of cerebellum)

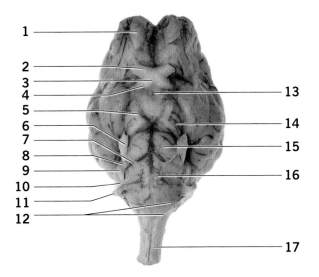

FIGURE 5.15 Sheep Brain with Cranial Nerves, Ventral View

1. Olfactory bulb
2. Optic nerve (II)
3. Optic chiasma
4. Optic tract
5. Oculomotor nerve (III)

6. Trigeminal nerve (V)
7. Abducens nerve (VI)
8. Facial nerve (VII)
9. Vestibulocochlear nerve (VIII)

10. Glossopharyngeal nerve (IX)
11. Vagus nerve (X)
12. Hypoglossal nerve (XII)
13. Infundibulum

14. Midbrain (cerebral peduncle)
15. Pons
16. Medulla oblongata (pyramid)
17. Spinal cord

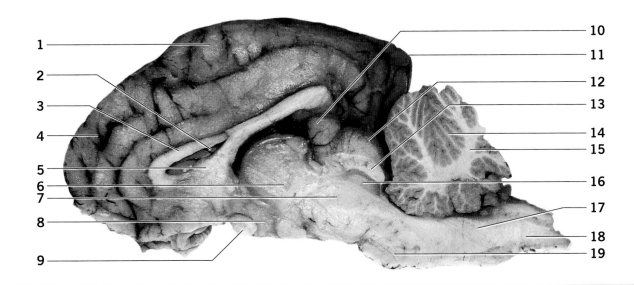

FIGURE 5.16 Sheep Brain, Midsagittal View

1. Parietal lobe
2. Lateral ventricle
3. Corpus callosum
4. Frontal lobe
5. Fornix

6. Thalamus
7. Midbrain
8. Hypothalamus
9. Optic chiasma
10. Pineal gland

11. Occipital lobe
12. Superior colliculus
13. Inferior colliculus
14. Cerebellum
15. Arbor vitae (of cerebellum)

16. Cerebral aqueduct
17. Medulla oblongata
18. Spinal cord
19. Pons

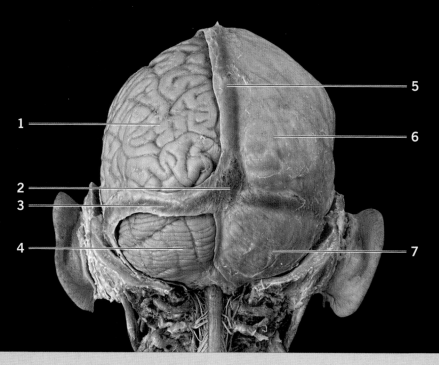

FIGURE 5.17 Brain with Dural Venous Sinuses, Posterior View

1. Left cerebral hemisphere
2. Confluence of sinuses
3. Transverse sinus
4. Left cerebellar hemisphere
5. Superior sagittal sinus
6. Dura mater covering right cerebral hemisphere
7. Dura mater covering right cerebellar hemisphere

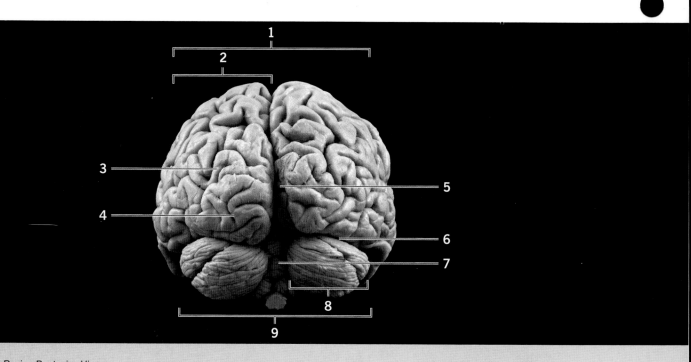

FIGURE 5.18 Brain, Posterior View

1. Cerebrum
2. Cerebral hemisphere
3. Parietal lobe
4. Occipital lobe
5. Longitudinal fissure
6. Transverse fissure
7. Vermis
8. Cerebellar hemisphere
9. Cerebellum

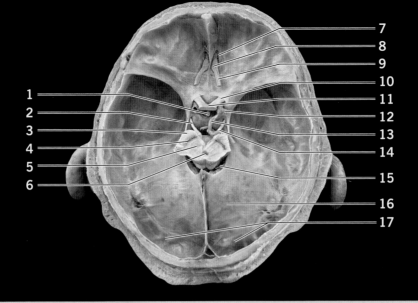

FIGURE 5.19 Floor of Cranial Cavity with Midbrain, Dura Mater, and Cranial Nerves

1. Infundibulum
2. Oculomotor nerve (III)
3. Crus cerebri
4. Substantia nigra
5. Red nucleus

6. Cerebral aqueduct
7. Olfactory nerves passing through cribriform plate
8. Olfactory bulb
9. Olfactory tract

10. Optic nerve
11. Optic chiasma
12. *Internal carotid artery*
13. *Posterior communicating artery*

14. *Posterior cerebral artery*
15. Trochlear nerve (IV)
16. Tentorium cerebelli
17. Transverse sinuses

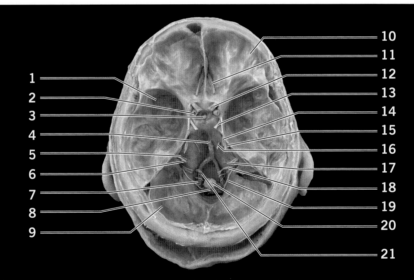

FIGURE 5.20 Floor of Cranial Cavity with Spinal Cord, Dura Mater, and Cranial Nerves

1. Middle cranial fossa
2. *Internal carotid artery*
3. Infundibulum
4. *Basilar artery*
5. Facial nerve (VII)

6. Vestibulocochlear nerve (VIII)
7. *Vertebral artery*
8. Spinal cord
9. Posterior cranial fossa
10. Anterior cranial fossa

11. Olfactory tract
12. Optic nerve (II)
13. Oculomotor nerve (III)
14. Trochlear nerve (IV)
15. Trigeminal nerve (V)
16. Abducens nerve (VI)

17. Glossopharyngeal nerve (IX)
18. Vagus nerve (X)
19. Accessory nerve (XI)
20. Medulla oblongata
21. Hypoglossal nerve (XII)

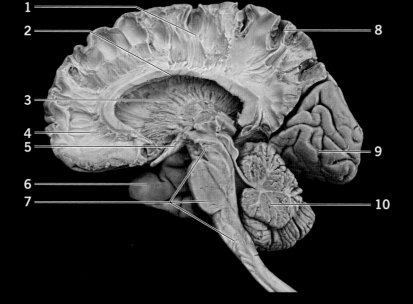

FIGURE 5.21 Brain, Parasagittal View

1. Association fibers	**3.** Internal capsule (projection fibers)	**5.** Optic tract	**8.** Cerebral cortex
2. Corpus callosum (commissural fibers)	**4.** Corona radiata (projection fibers)	**6.** Temporal lobe	**9.** Occipital lobe
		7. Brain stem	**10.** Cerebellum

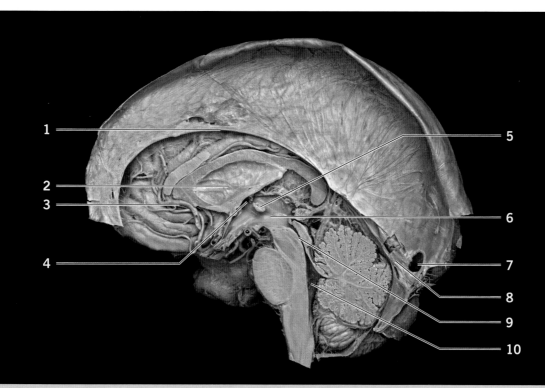

FIGURE 5.22 Brain with Dura Mater, Midsagittal View

1. Inferior sagittal sinus	**4.** Interventricular foramen	**6.** Third ventricle	**8.** Straight sinus
2. Septum pellucidum	**5.** Interthalamic adhesion	**7.** Confluence of the sinuses	**9.** Cerebral aqueduct
3. *Anterior cerebral artery*			**10.** Fourth ventricle

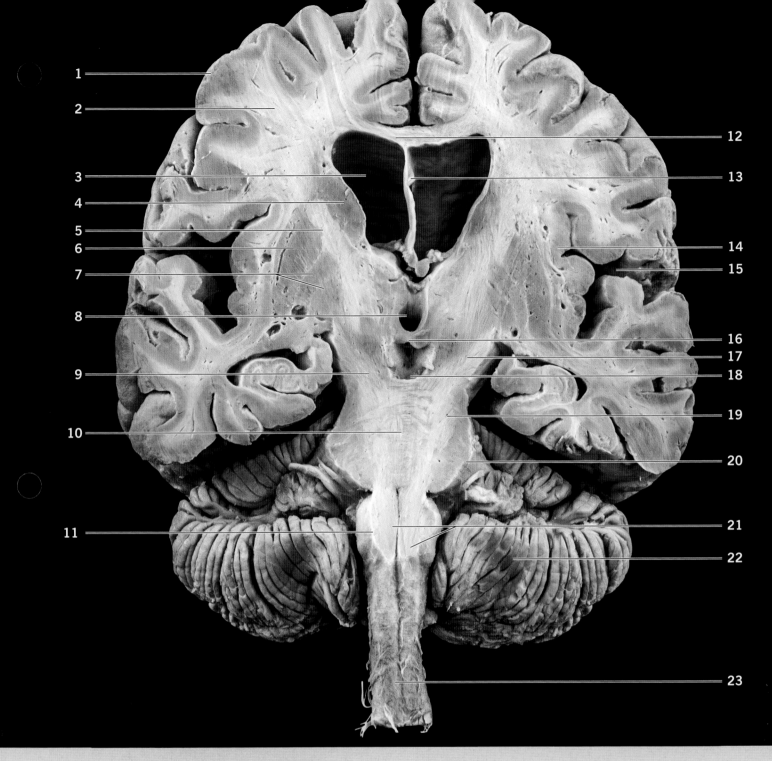

FIGURE 5.23 Brain, Frontal View

1. Cerebral cortex
2. White matter
3. Lateral ventricle
4. Caudate nucleus
5. Internal capsule
6. Putamen
7. Globus pallidus

8. Interventricular foramen
9. Midbrain
10. Pons
11. Medulla oblongata
12. Corpus callosum
13. Septum pellucidum

14. Insula
15. Lateral sulcus
16. Interthalamic adhesion
17. Crus cerebri
18. Cerebral aqueduct
19. Fibers of pyramidal tracts

20. Middle cerebellar peduncle
21. Pyramids
22. Cerebellum
23. Spinal cord

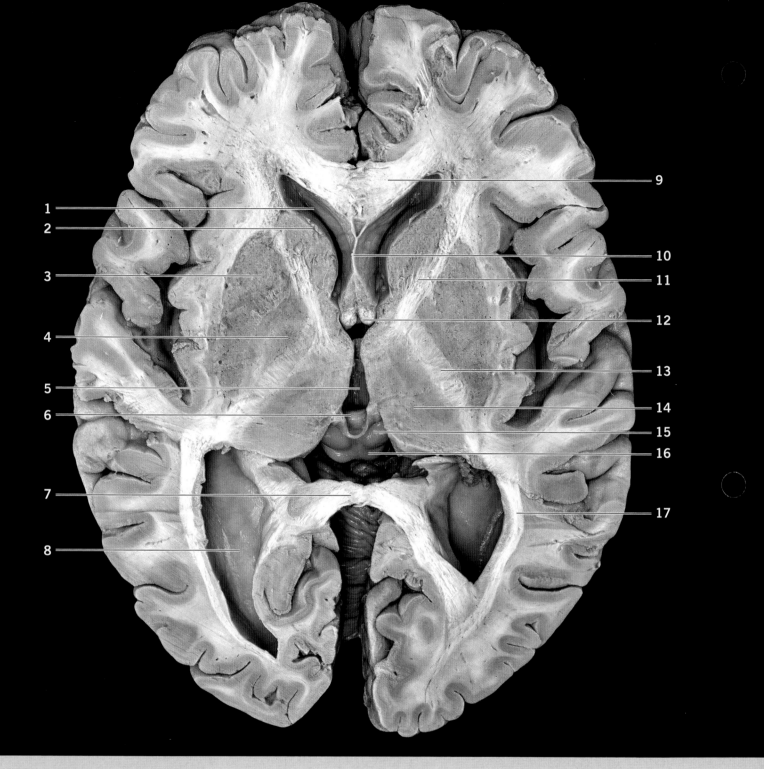

FIGURE 5.24 Brain, Transverse View

1. Lateral ventricle (anterior horn)
2. Caudate nucleus
3. Putamen
4. Globus pallidus
5. Third ventricle
6. Pineal gland
7. Corpus callosum (splenium)
8. Lateral ventricle (inferior horn)
9. Corpus callosum (genu)
10. Septum pellucidum
11. Internal capsule (anterior limb)
12. Fornix
13. Internal capsule (posterior limb)
14. Thalamus
15. Superior colliculus
16. Inferior colliculus
17. Optic radiations

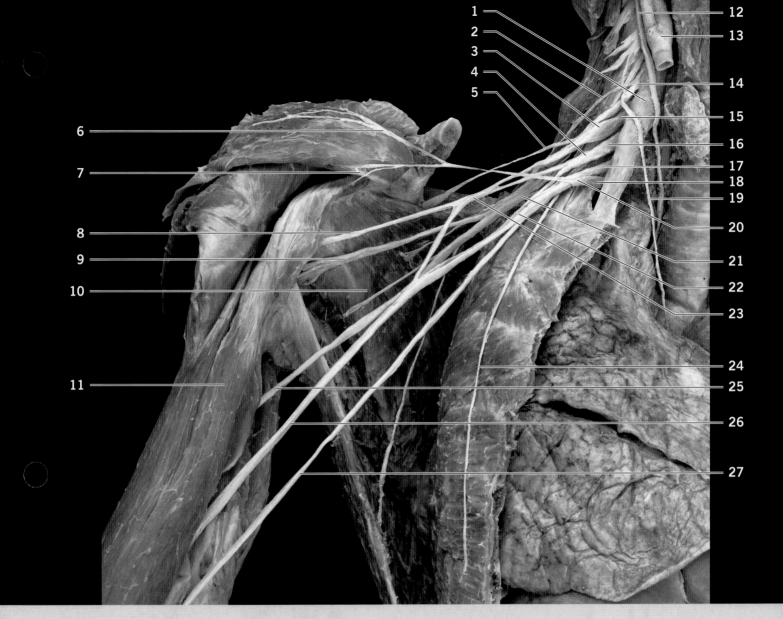

FIGURE 5.25 Brachial Plexus

1. *Anterior scalene muscle*
2. Dorsal scapular nerve
3. Upper trunk
4. Middle trunk
5. Suprascapular nerve
6. Middle pectoral nerve
7. Lateral pectoral nerve
8. Musculocutaneous nerve
9. Axillary nerve
10. *Subscapularis muscle*
11. *Short head of biceps brachii muscle*
12. Vagus nerve
13. *Common carotid artery*
14. C5
15. C6
16. C7
17. C8
18. T1
19. Phrenic nerve
20. Lower trunk
21. Posterior cord
22. Medial cord
23. Lateral cord
24. Long thoracic nerve
25. Radial nerve
26. Median nerve
27. Ulnar nerve

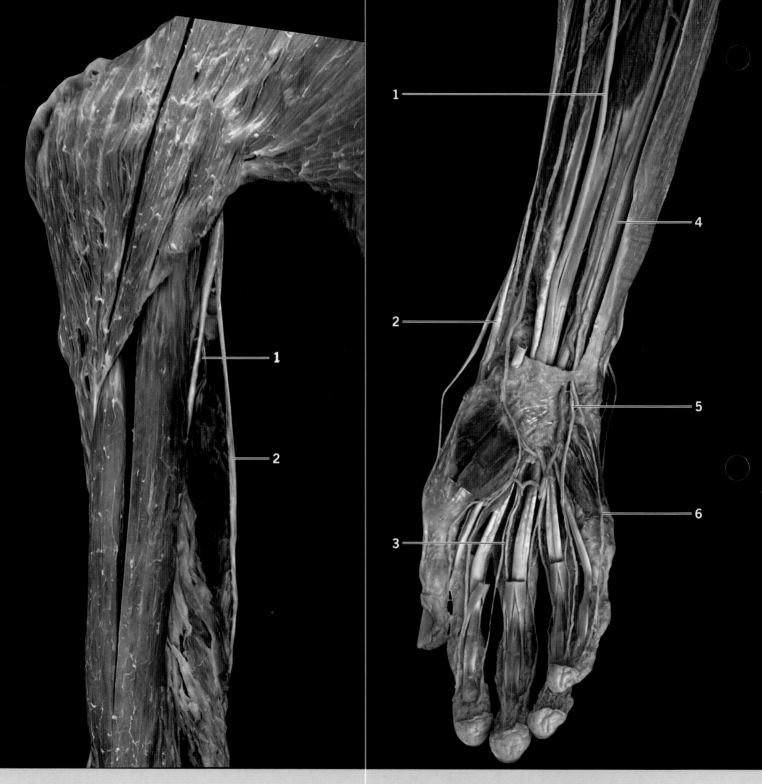

FIGURE 5.26 Nerves of the Arm, Anterior View

FIGURE 5.27 Deep Nerves of the Forearm and Hand, Anterior View

1. Median nerve
2. Ulnar nerve

1. Median nerve
2. Radial nerve
3. Palmar digital branches of median nerve

4. Ulnar nerve
5. Superficial branch of ulnar nerve
6. Palmar digital branch of ulnar nerve

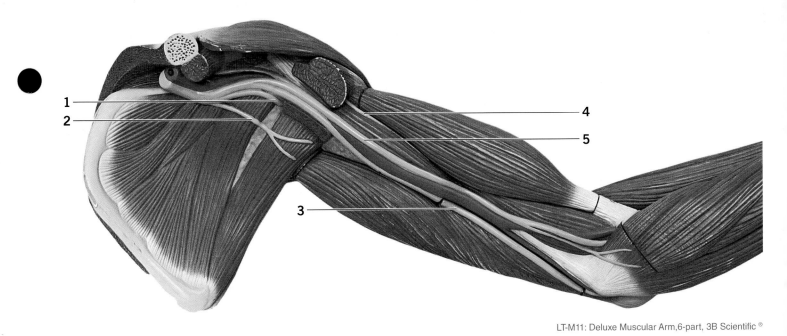

LT-M11: Deluxe Muscular Arm,6-part, 3B Scientific ®

FIGURE 5.28 Nerves of the Arm, Anterior View

1. Radial nerve
2. Axillary nerve

3. Ulnar nerve
4. Musculocutaneous nerve

5. Median nerve

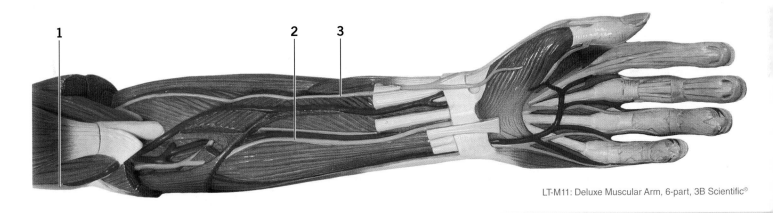

LT-M11: Deluxe Muscular Arm, 6-part, 3B Scientific®

FIGURE 5.29 Deep Nerves of the Forearm and Hand, Anterior View

1. Ulnar nerve

2. Median nerve

3. Radial Nerve (superficial branch)

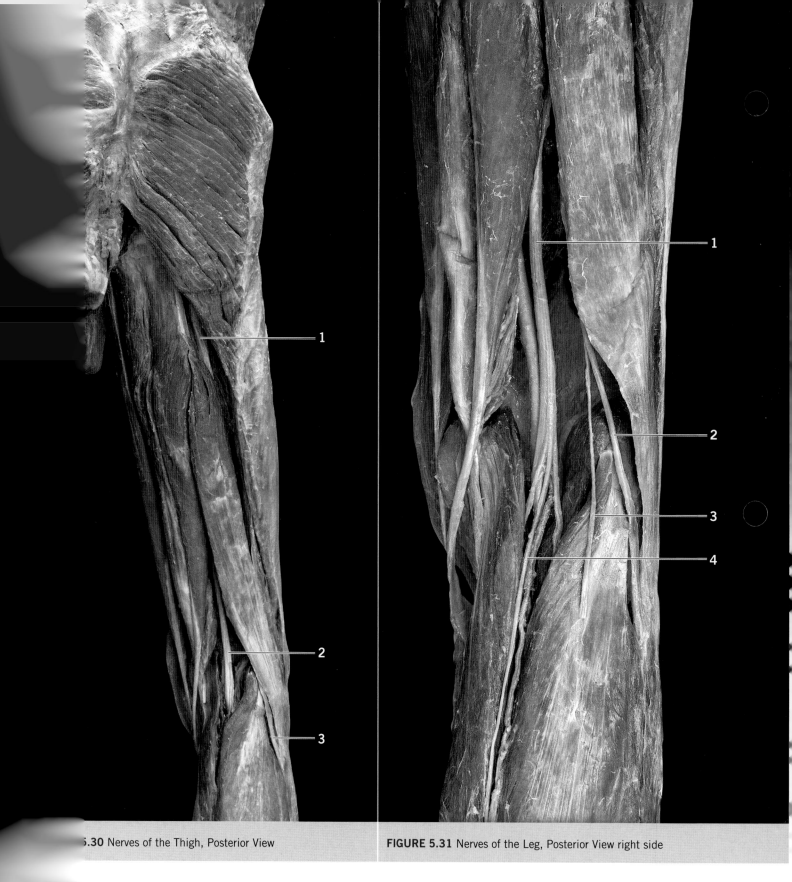

FIGURE 5.31 Nerves of the Leg, Posterior View right side

atic nerve

ial nerve

mmon fibular nerve

1. Tibial nerve

2. Common fibular nerve

3. Lateral sural cutaneous nerve

4. Medial sural cutaneous nerve

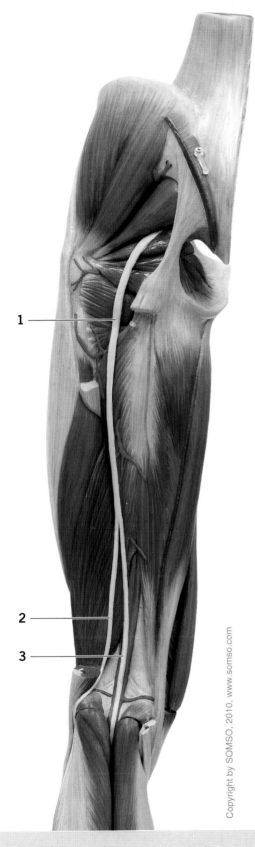

FIGURE 5.32 Nerves of the Thigh, Posterior View

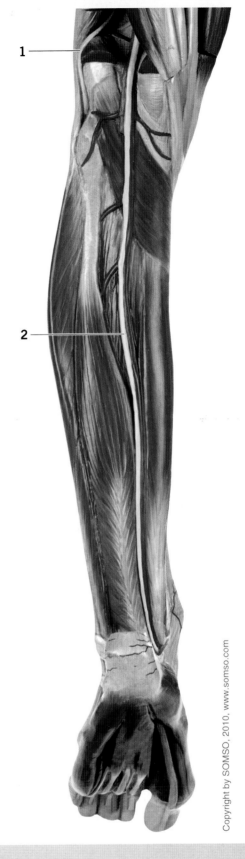

FIGURE 5.33 Nerves of the Leg, Posterior View

Copyright by SOMSO, 2010, www.somso.com

1. Sciatic nerve
2. Common fibular nerve
3. Tibial nerve

1. Common fibular nerve
2. Tibial nerve

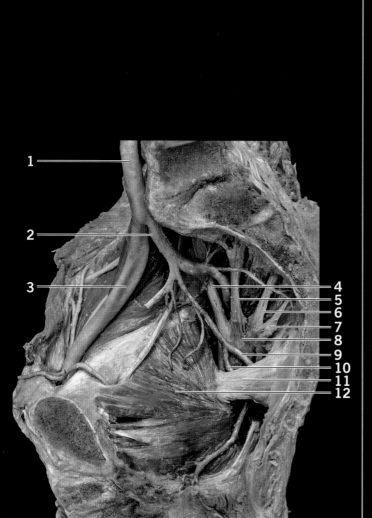

FIGURE 5.34 Sacral Plexus, Anterolateral View

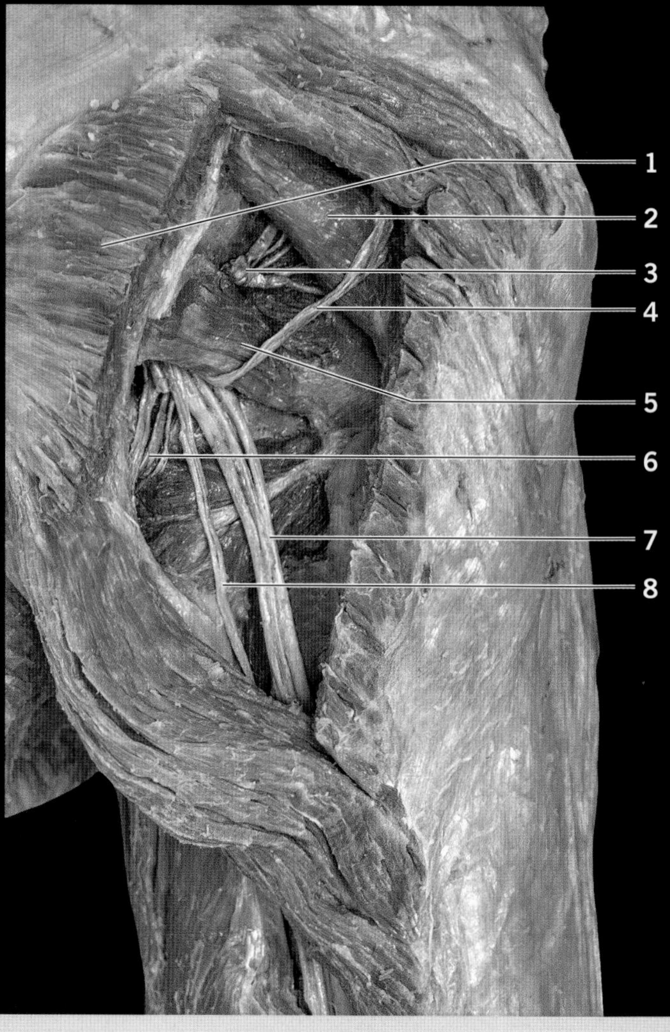

FIGURE 5.35 Nerves of the Buttock, Posterior View

1. *Common iliac artery*
2. *Internal iliac artery*
3. *External iliac artery*
4. Lumbosacral trunk
5. S1
6. S2
7. S3

8. Sacral plexus
9. *Inferior gluteal artery*
10. *Internal pudendal artery*
11. *Sacrospinous ligament*
12. *Obturator internus muscle*

1. *Gluteus maximus muscle, cut*
2. *Gluteus minimus muscle*
3. Superior gluteal nerve
4. Inferior gluteal nerve

5. *Piriformis muscle*
6. Pudendal nerve
7. Sciatic nerve
8. Posterior femoral cutaneous nerve

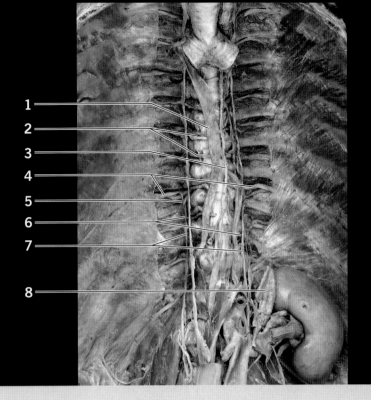

FIGURE 5.36 Sympathetic Trunk, Sympathetic Trunk Ganglia, and Adrenal Gland, Anterior View

1. *Azygos vein*
2. Posterior intercostal vessels
3. *Thoracic duct*
4. Intercostal nerves
5. Ramus communicans
6. Sympathetic trunks
7. Sympathetic trunk ganglia
8. *Adrenal gland*

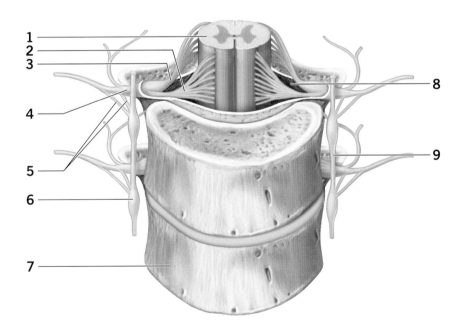

FIGURE 5.37 Diagram of the Spinal Cord, Sympathetic Trunk, and Sympathetic Trunk Ganglia, Anterolateral View

1. Spinal cord
2. Ventral root
3. Dorsal root
4. Spinal nerve
5. Rami communicans
6. Sympathetic trunk ganglion
7. *Body of vertebra*
8. Intervertebral foramen
9. Sympathetic trunk

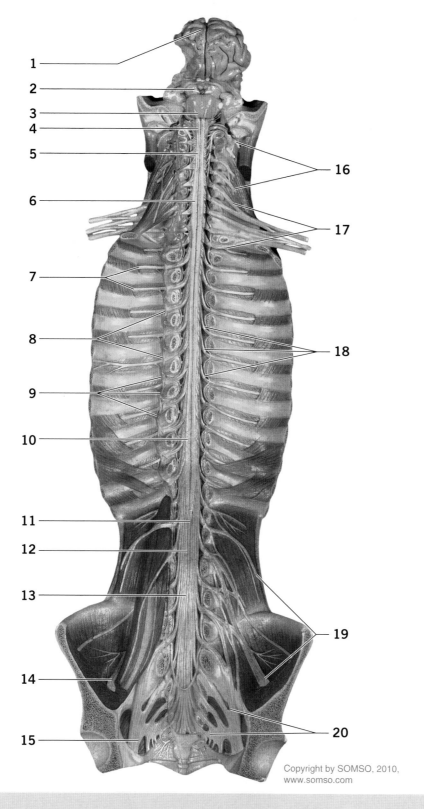

FIGURE 5.38 Central, Peripheral, and Autonomic Nervous Systems

1. Cerebrum
2. Midbrain
3. Pons
4. Medulla oblongata
5. Spinal cord
6. Cervical enlargement

7. Intercostal nerves
8. Sympathetic trunk
9. Sympathetic trunk ganglia
10. Lumbar enlargement
11. Conus medullaris

12. Filum terminale
13. Cauda equina
14. Femoral nerve
15. Sciatic nerve
16. Cervical plexus
17. Brachial plexus

18. Spinal nerves
19. Lumbar plexus
20. Sacral plexus

Autonomic Nervous System

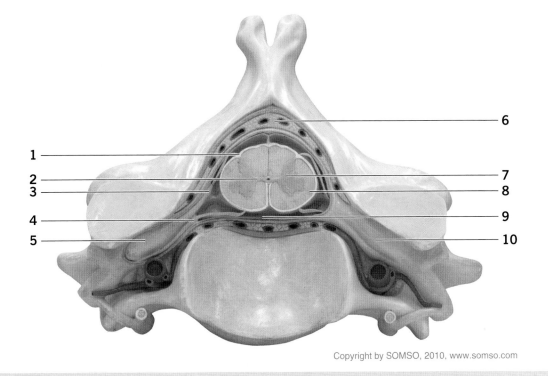

Copyright by SOMSO, 2010, www.somso.com

FIGURE 5.39 Spinal Cord in Vertebral Canal

1. Pia mater	**4.** Ventral root	**7.** Gray matter	**10.** Dura mater
2. Gray commissure	**5.** Dorsal root ganglion	**8.** White matter	
3. Dorsal root	**6.** Epidural space	**9.** Arachnoid mater	

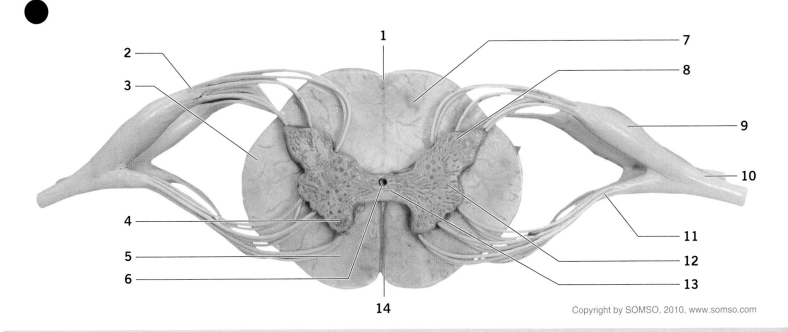

Copyright by SOMSO, 2010, www.somso.com

FIGURE 5.40 Spinal Cord and Spinal Nerve Roots

1. Dorsal median sulcus	**5.** Ventral funiculus	**9.** Dorsal root ganglion	**13.** Gray commissure
2. Dorsal root	**6.** Central canal	**10.** Spinal nerve	**14.** Ventral median fissure
3. Lateral funiculus	**7.** Dorsal funiculus	**11.** Ventral root	
4. Ventral horn	**8.** Dorsal horn	**12.** Lateral horn	

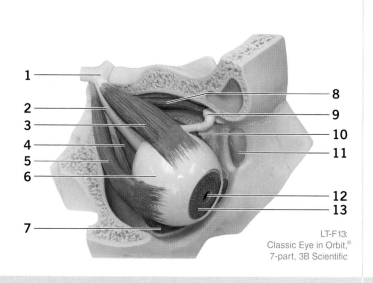

LT-F13:
Classic Eye in Orbit,®
7-part, 3B Scientific

FIGURE 5.41 Extrinsic Muscles of the Eye

1. Optic chiasma
2. Optic nerve
3. Superior rectus muscle
4. Inferior rectus muscle
5. Lateral rectus muscle
6. Sclera
7. Inferior oblique muscle
8. Superior oblique muscle
9. Trochlea
10. Medial rectus muscle
11. Lacrimal sac
12. Pupil
13. Iris

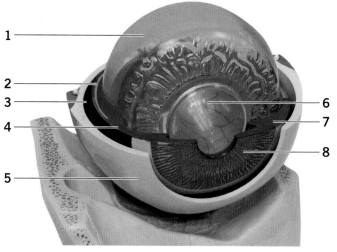

LT-F12: Giant Eye with Eyelid and Lachrymal System, 8-part, 3B Scientific®

FIGURE 5.42 Internal Structures of the Eye

1. Vitreous humor
2. Retina
3. Choroid
4. Ciliary body
5. Sclera
6. Lens
7. Ciliary muscle
8. Iris

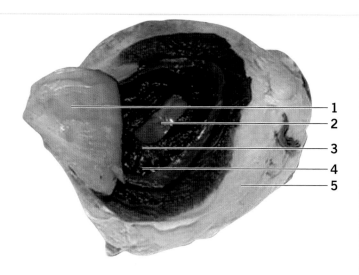

FIGURE 5.43 Cow Eye, Whole with Corneal Flap

1. Cornea
2. Pupil
3. Iris
4. Ciliary muscle
5. Sclera

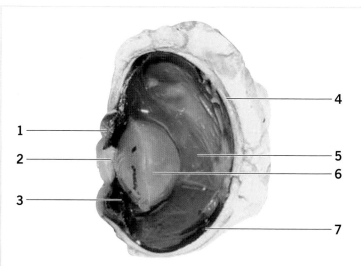

FIGURE 5.44 Cow Eye, Midsagittal Section

1. Cornea
2. Pupil
3. Iris
4. Sclera
5. Vitreous humor in posterior cavity
6. Lens
7. Choroid

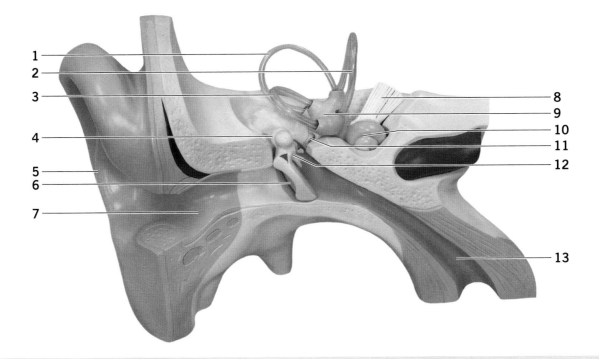

FIGURE 5.45 Ear, Anterior View

1. Posterior semicircular duct
2. Anterior semicircular duct
3. Lateral semicircular duct
4. Incus
5. Auricle
6. Tympanic membrane
7. External acoustic meatus
8. Vestibulocochlear nerve
9. Vestibule
10. Cochlea
11. Stapes
12. Malleus
13. Pharyngotympanic tube

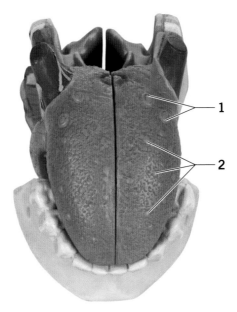

FIGURE 5.46 Papillae of Tongue, Superior View

1. Vallate papillae
2. Fungiform papillae

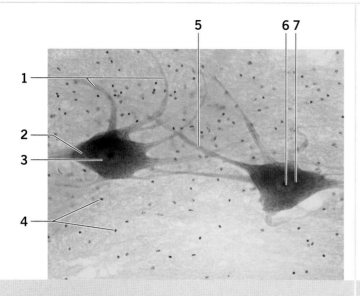

FIGURE 5.47 Motor Neuron Smear, 400×

1. Dendrites
2. Axon hillock
3. Nucleus
4. Neuroglia
5. Axon
6. Nucleolus
7. Cell body

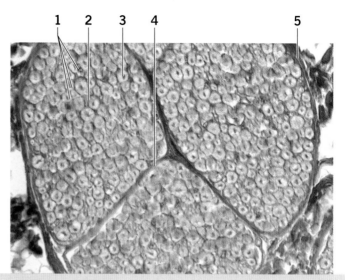

FIGURE 5.48 Peripheral Nerve, Cross Section, 400×

1. Myelinated axons
2. Endoneurium
3. Myelin sheath
4. Perineurium
5. Epineurium

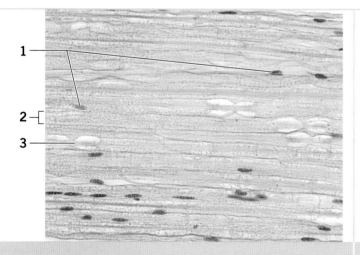

FIGURE 5.49 Peripheral Nerve, Longitudinal Section, 600×

1. Schwann cell nuclei
2. Myelinated nerve fiber
3. Node of Ranvier

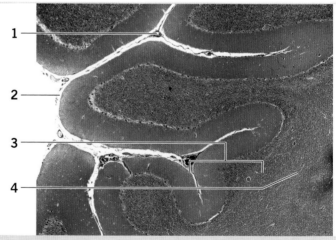

FIGURE 5.50 Cerebellum, Sagittal Section, 40×

1. Sulcus
2. Gyrus
3. Gray matter
4. White matter

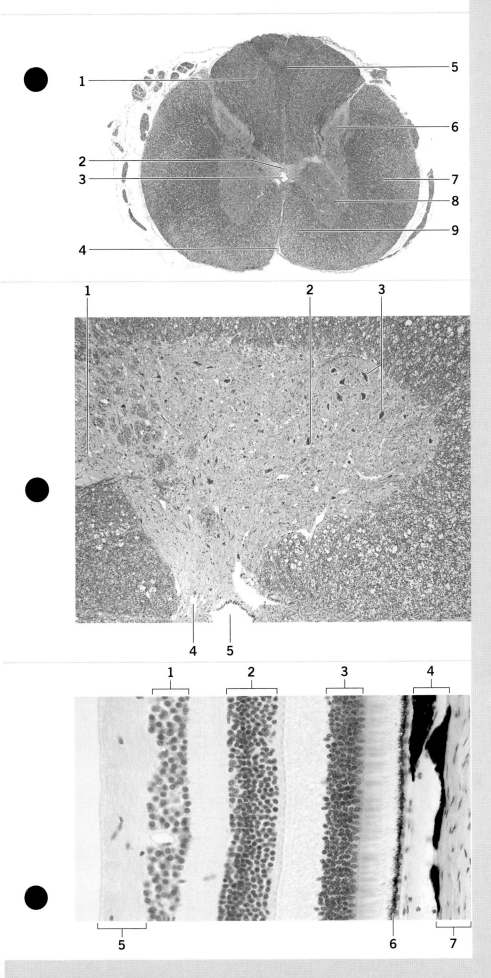

FIGURE 5.51 Spinal Cord, Cervical Region, Cross Section, 20×

1. Dorsal funiculus
2. Gray commissure
3. Central canal
4. Ventral median fissure
5. Dorsal median sulcus
6. Dorsal horn
7. Lateral funiculus
8. Ventral horn
9. Ventral funiculus

FIGURE 5.52 Spinal Cord, Cervical Region, Cross Section, 100×

1. Dorsal horn
2. Ventral horn
3. Cell bodies of somatic motor neurons
4. Gray commissure
5. Central canal

FIGURE 5.53 Retina, Cross Section, 400×

1. Nuclei of ganglion cells
2. Nuclei of bipolar cells
3. Nuclei of rods and cones
4. Choroid
5. Axons of ganglion cells
6. Pigmented layer
7. Sclera

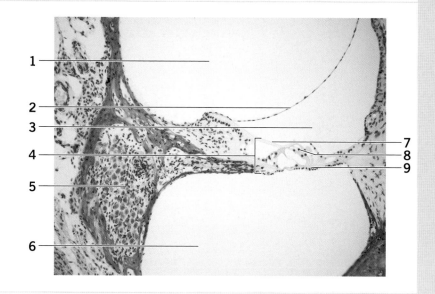

FIGURE 5.54 Cochlea, Cross Section, 200×

1. Scala vestibuli
2. Vestibular membrane
3. Scala media/cochlear duct
4. Spiral organ
5. Spiral ganglion
6. Scala tympani
7. Tectorial membrane
8. Outer hair cells
9. Basilar membrane

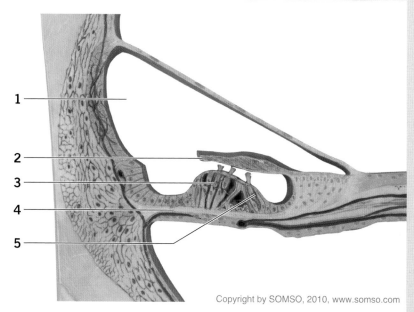

Copyright by SOMSO, 2010, www.somso.com

FIGURE 5.55 Cochlea, Spiral Organ

1. Scala media/cochlear duct
2. Tectorial membrane
3. Outer hair cells
4. Basilar membrane
5. Inner hair cell

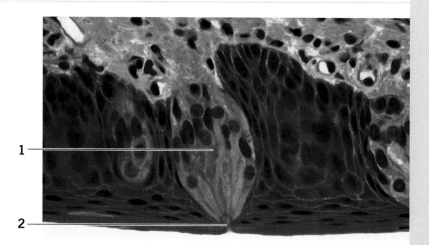

FIGURE 5.56 Taste Buds, Cross Section, 200×

1. Taste bud
2. Taste pore

Tissues of the Nervous System

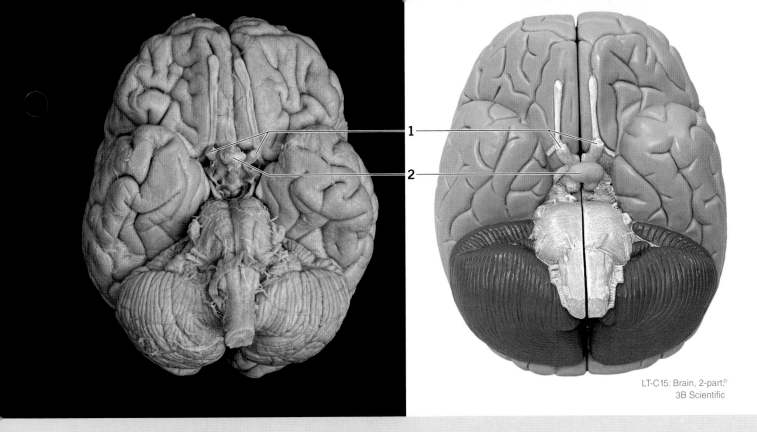

LT-C15: Brain, 2-part.®
3B Scientific

FIGURE 6.1 Brain, Inferior View

1. *Optic nerves*   2. Pituitary gland

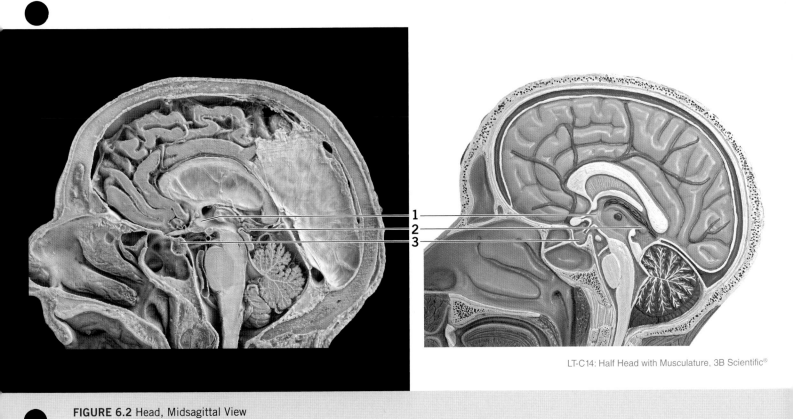

LT-C14: Half Head with Musculature, 3B Scientific®

FIGURE 6.2 Head, Midsagittal View

1. Hypothalamus   2. Pineal gland   3. Pituitary gland

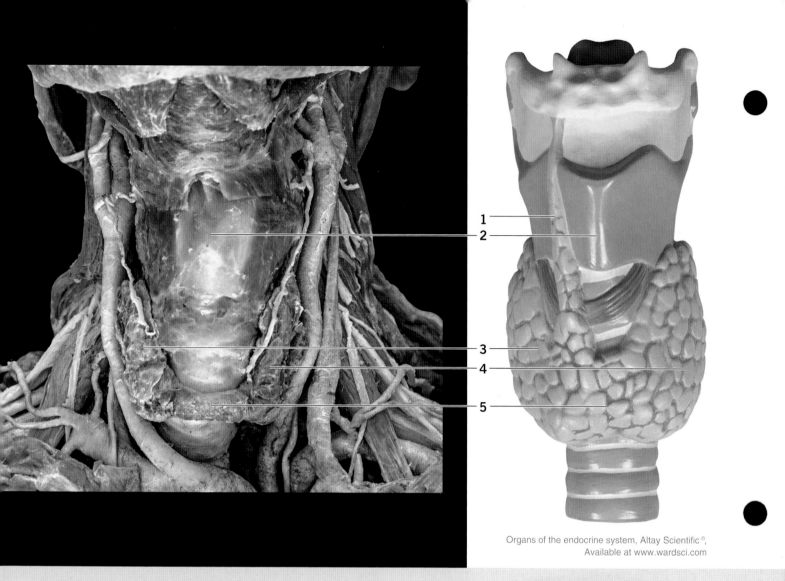

FIGURE 6.3 Thyroid Gland, Anterior View

1. Pyramidal lobe of thyroid gland
2. *Thyroid cartilage*

3. Right lateral lobe of thyroid gland
4. Left lateral lobe of thyroid gland

5. Isthmus

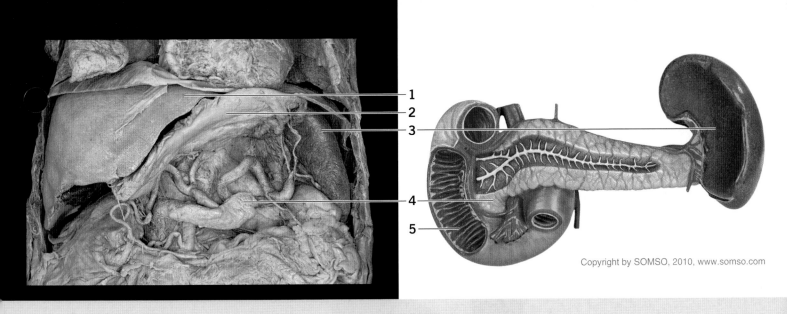

FIGURE 6.4 Upper Abdomen, Anterior View

1. Liver
2. Stomach

3. Spleen
4. Pancreas

5. Duodenum

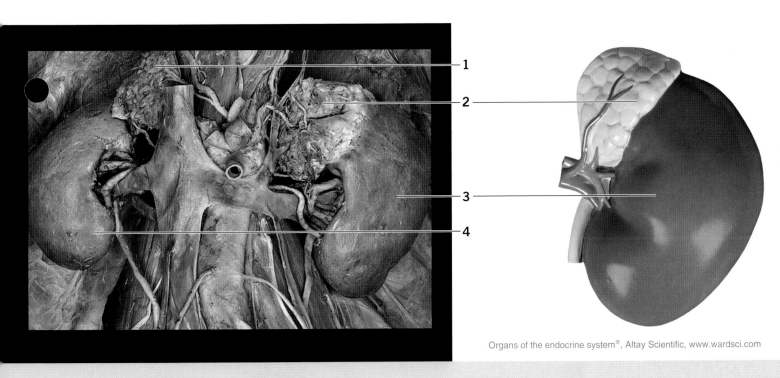

FIGURE 6.5 Adrenal Glands and Kidneys, Anterior View

1. Right adrenal gland

2. Left adrenal gland

3. Left kidney

4. Right kidney

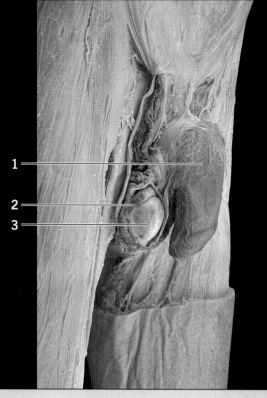

FIGURE 6.6 Testis, Lateral View

1. *Penis* **2.** *Epididymis* **3.** Testis

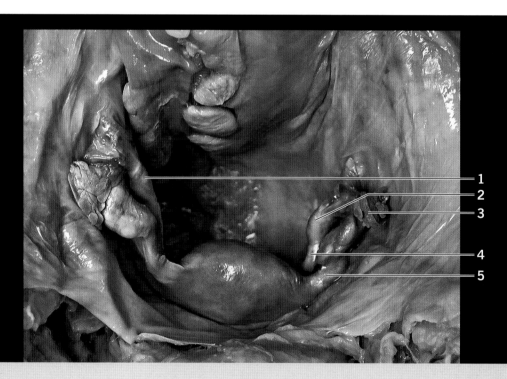

FIGURE 6.7 Ovaries, Superior View

1. Right ovary **3.** *Left fimbriae* **5.** *Left uterine tube*
2. Left ovary **4.** *Left ovarian ligament*

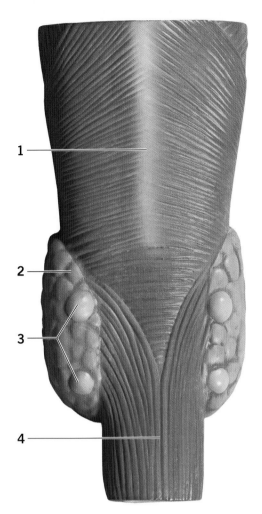

1

2

3

4

FIGURE 6.8 Thyroid and Parathyroid Glands, Posterior View

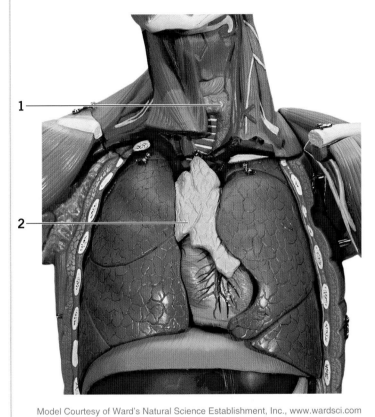

1

2

FIGURE 6.9 Thymus and Thyroid Glands, Anterior View

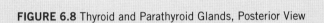

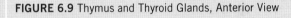

1. *Inferior pharyngeal constrictor muscle*
2. Thyroid gland
3. Parathyroid glands
4. *Esophagus*

1. Thyroid gland
2. Thymus

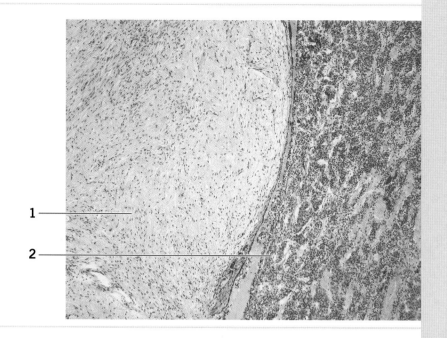

FIGURE 6.10 Pituitary Gland,
Cross Section 100×

1. Posterior pituitary
2. Anterior pituitary

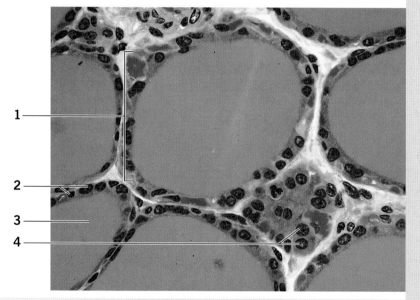

FIGURE 6.11 Thyroid Gland,
Cross Section 1000×

1. Follicle
2. Follicular cells
3. Colloid
4. Parafollicular (C) cells

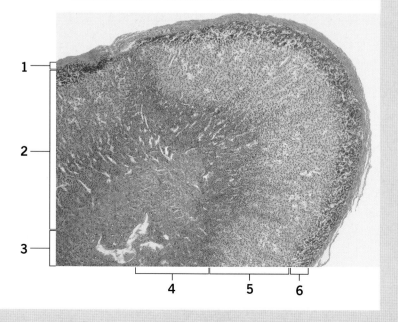

FIGURE 6.12 Adrenal Gland,
Cross Section 100×

1. Capsule
2. Cortex
3. Medulla
4. Zona reticularis
5. Zona fasciculata
6. Zona glomerulosa

Tissues of the Endocrine System

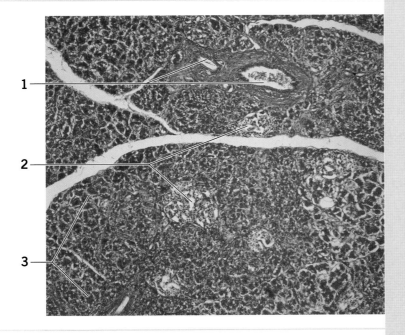

FIGURE 6.13 Pancreas, Cross Section 100×

1. Ducts
2. Pancreatic islets
3. Acinar tissue

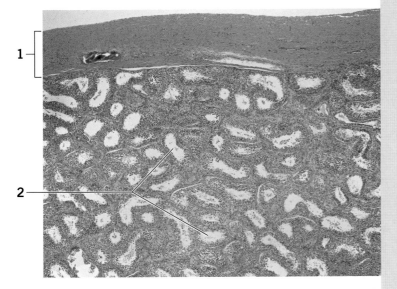

FIGURE 6.14 Testis, Cross Section 40×

1. Tunica albuginea
2. Seminiferous tubules

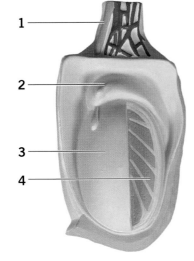

Organs of the Endocrine System, Altay Scientific®,
Available at www.wardsci.com

FIGURE 6.15 Testis, Anterior View

1. *Ductus deferens*
2. *Epididymis*
3. Testis
4. Lobule with seminiferous tubules

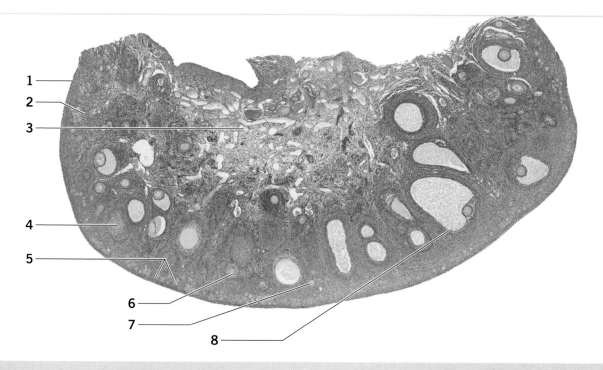

FIGURE 6.16 Ovary, Sagittal Section 40×

1. Tunica albuginea	**3.** Medulla	**5.** Primordial follicles	**7.** Primary follicle
2. Cortex	**4.** Corpus luteum	**6.** Secondary follicle	**8.** Mature tertiary follicle

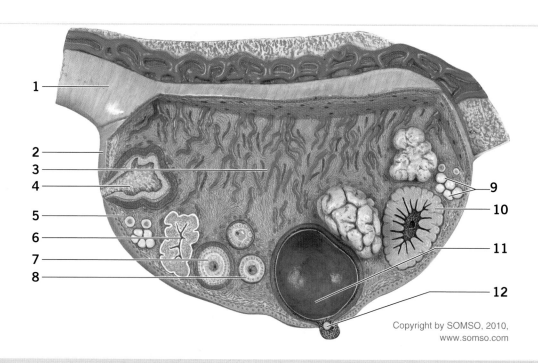

Copyright by SOMSO, 2010,
www.somso.com

FIGURE 6.17 Ovary, Bisected View

1. Ovarian ligament	**4.** Ruptured follicle	**7.** Secondary follicle	**10.** Corpus luteum
2. Tunica albuginea	**5.** Cortex	**8.** Primary follicle	**11.** Mature tertiary follicle
3. Medulla	**6.** Corpus albicans	**9.** Primordial follicles	**12.** Oocyte

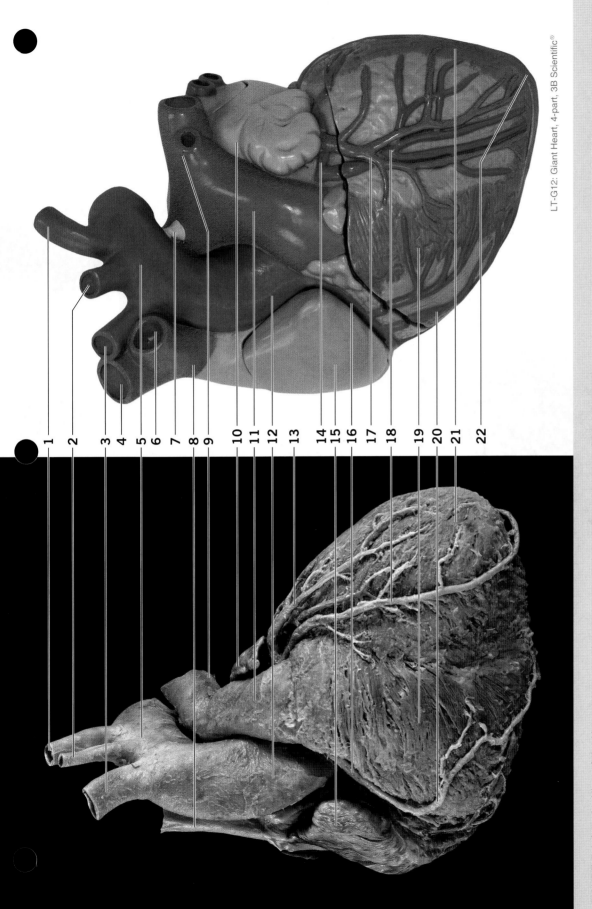

LT-G12: Giant Heart, 4-part, 3B Scientific®

FIGURE 7.1 Anterior View of the Heart

1. Left subclavian artery
2. Left common carotid artery
3. Brachiocephalic trunk
4. Right brachiocephalic vein
5. Aortic arch
6. Left brachiocephalic vein

7. *Ligamentum arteriosum*
8. Superior vena cava
9. Left pulmonary artery
10. Left auricle
11. Pulmonary trunk
12. Ascending aorta

13. Left marginal artery
14. Circumflex artery
15. Right auricle
16. Right coronary artery
17. Great cardiac vein
18. Anterior interventricular artery

19. Right ventricle
20. Right marginal artery
21. Left ventricle
22. Apex

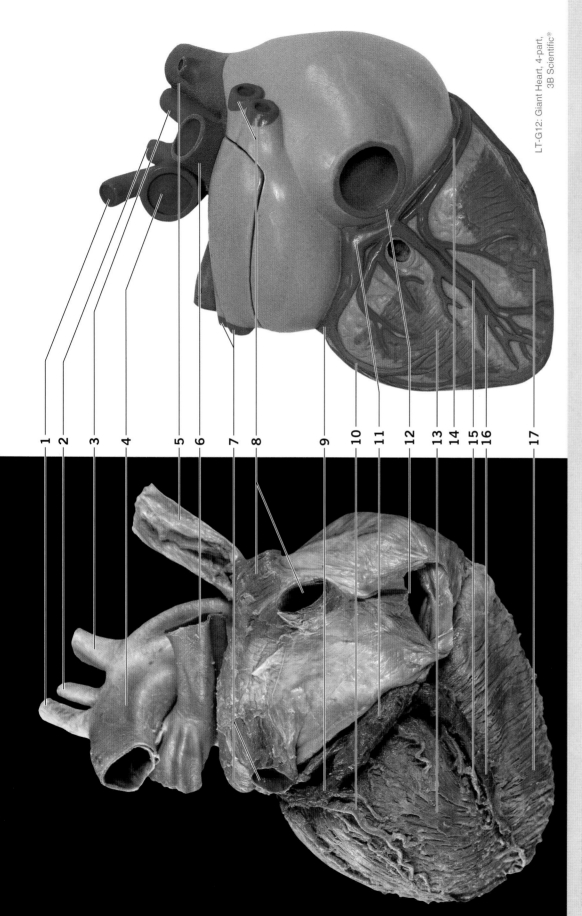

FIGURE 7.2 Posterior View of the Heart

1. Left subclavian artery
2. Left common carotid artery
3. Brachiocephalic trunk
4. Aortic arch
5. Superior vena cava
6. Right pulmonary artery
7. Left pulmonary veins
8. Right pulmonary veins
9. Great cardiac vein
10. Posterior vein of left ventricle
11. Coronary sinus
12. Inferior vena cava
13. Left ventricle
14. Small cardiac vein
15. Posterior interventricular artery
16. Middle cardiac vein
17. Right ventricle

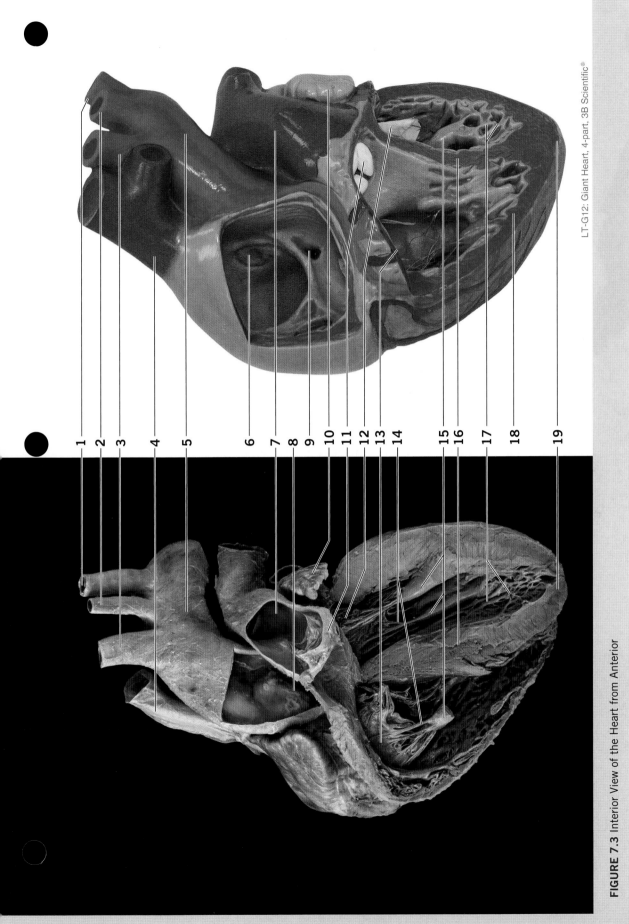

LT-G12: Giant Heart, 4-part, 3B Scientific®

FIGURE 7.3 Interior View of the Heart from Anterior

1. Left subclavian artery
2. Left common carotid artery
3. Brachiocephalic trunk
4. Superior vena cava
5. Aortic arch
6. Fossa ovalis

7. Pulmonary trunk
8. Aortic semilunar valve
9. Ostium of coronary sinus
10. Left auricle
11. Pulmonary semilunar valve

12. Left atrioventricular (biscuspid/mitral) valve
13. Right atrioventricular (tricuspid) valve
14. Chordae tendineae
15. Papillary muscles

16. Interventricular septum
17. Trabeculae carneae of left ventricle
18. Right ventricle
19. Apex

Chapter 7 The Cardiovascular System

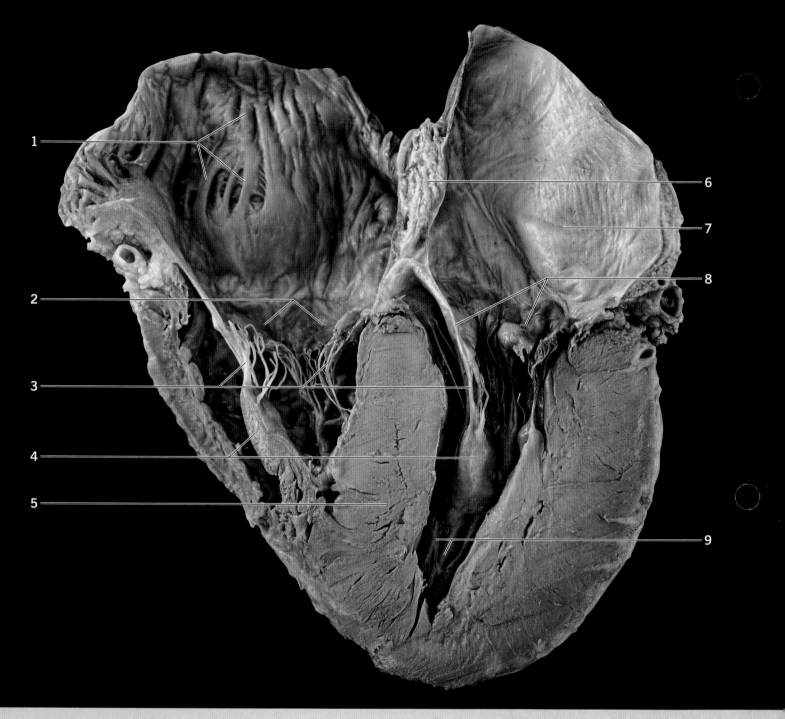

FIGURE 7.4 Bisected Heart

1. Pectinate muscles of right atrium
2. Tricuspid valve
3. Chordae tendineae
4. Papillary muscles
5. Interventricular septum
6. Interatrial septum
7. Left atrium
8. Bicuspid valve
9. Trabeculae carneae of left ventricle

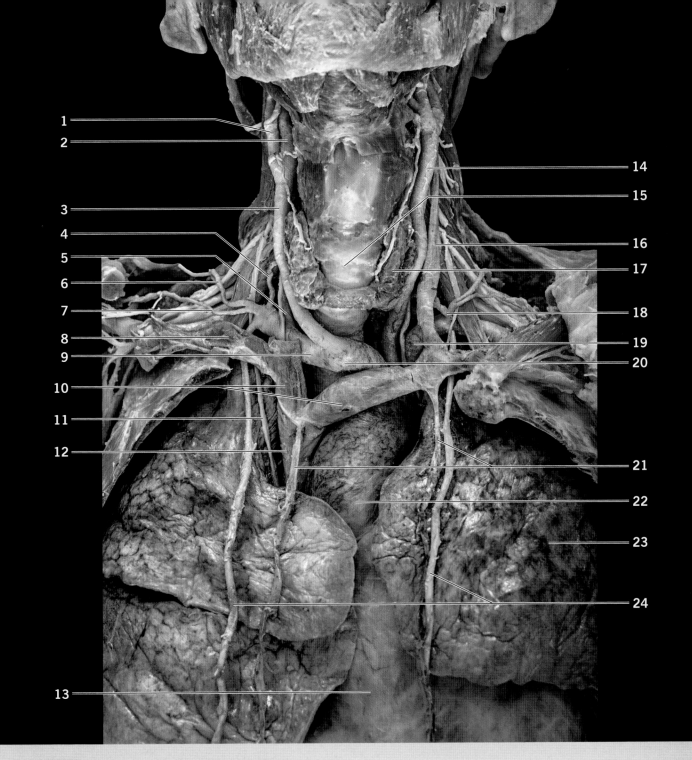

FIGURE 7.5 Blood Vessels of the Upper Thorax and Neck

1. External carotid artery
2. Internal carotid artery
3. Right common carotid artery
4. Inferior thyroid artery
5. Vertebral artery
6. Transverse cervical artery
7. Suprascapular artery
8. Subclavian vein
9. Right subclavian artery
10. Brachiocephalic veins
11. *Phrenic nerve*
12. Superior vena cava
13. Pericardial sac
14. Left common carotid artery
15. *Trachea*
16. Internal jugular vein
17. *Thyroid gland, left lobe*
18. Thyrocervical trunk
19. Left subclavian artery
20. Brachiocephalic trunk
21. Internal thoracic veins
22. Ascending aorta
23. *Left Lung*
24. Internal thoracic arteries

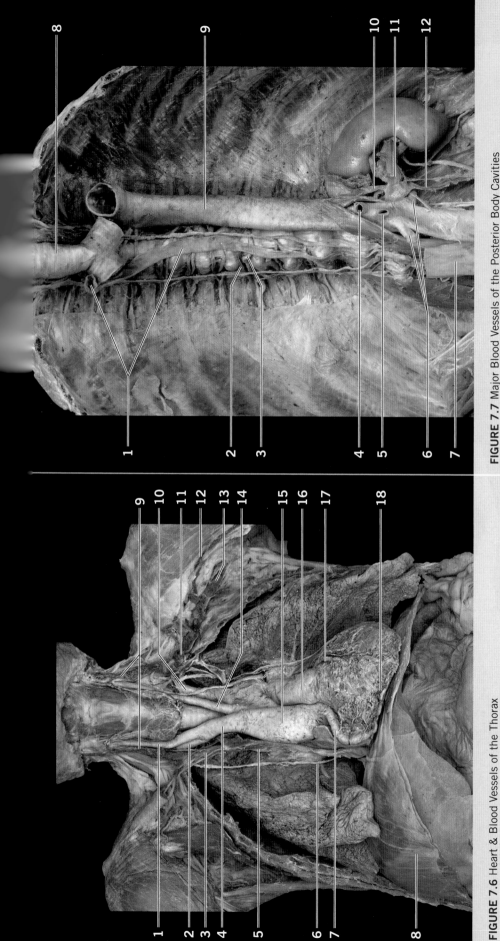

FIGURE 7.6 Heart & Blood Vessels of the Thorax

1. Right common carotid artery
2. Right subclavian artery
3. Right brachiocephalic vein
4. Brachiocephalic trunk
5. Superior vena cava
6. *Phrenic nerve*
7. Right coronary artery
8. Respiratory diaphragm
9. Internal jugular veins
10. Left subclavian artery
11. Subclavian vein
12. Cephalic vein
13. Axillary vein
14. Left common carotid artery
15. Aorta
16. Pulmonary trunk
17. Anterior interventricular artery
18. Right marginal artery

FIGURE 7.7 Major Blood Vessels of the Posterior Body Cavities

1. Azygous vein
2. Posterior intercostal vein
3. Posterior intercostal artery
4. Celiac trunk
5. Superior mesenteric artery
6. Renal arteries
7. Inferior vena cava
8. *Trachea*
9. Aorta
10. Suprarenal vein
11. Renal vein
12. Gonadal vein

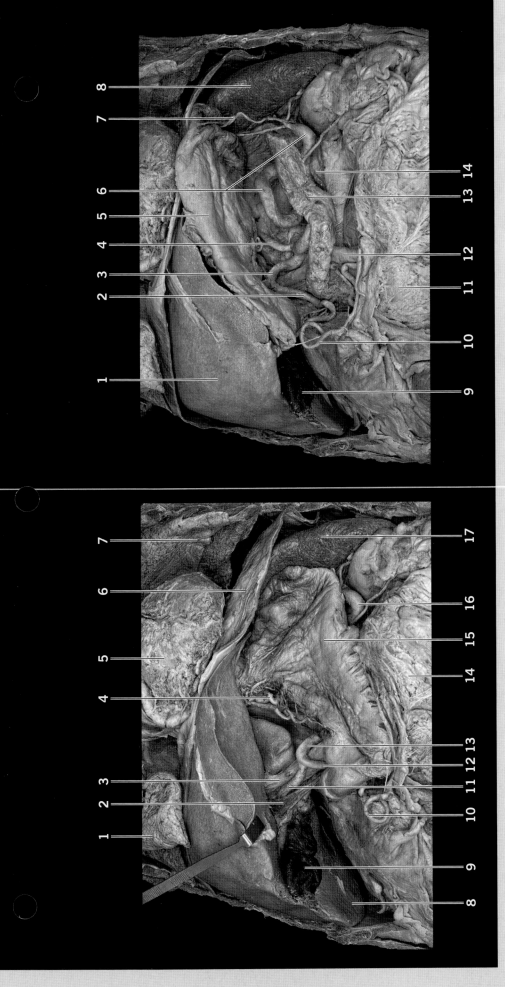

FIGURE 7.8 Major Vessels of the Superior Abdominal Cavity

1. *Right lung*
2. *Common hepatic duct*
3. *Hepatic portal vein*
4. *Left gastric artery*
5. *Heart*
6. *Respiratory diaphragm*
7. *Left lung*
8. *Liver*
9. *Gallbladder*
10. Right gastroepiploic artery
11. Proper hepatic artery
12. Gastroduodenal artery
13. Common hepatic artery
14. Greater omentum
15. Stomach
16. Splenic artery
17. Spleen

FIGURE 7.9 Branches of the Celiac Trunk and Superior Mesenteric Artery

1. *Liver*
2. Gastroduodenal artery
3. Common hepatic artery
4. Left gastric artery
5. *Stomach, reflected*
6. Splenic artery
7. Left gastroepiploic artery
8. *Spleen*
9. *Gallbladder*
10. Right gastroepiploic artery
11. *Greater omentum*
12. Superior mesenteric artery
13. *Pancreas*
14. Splenic vein

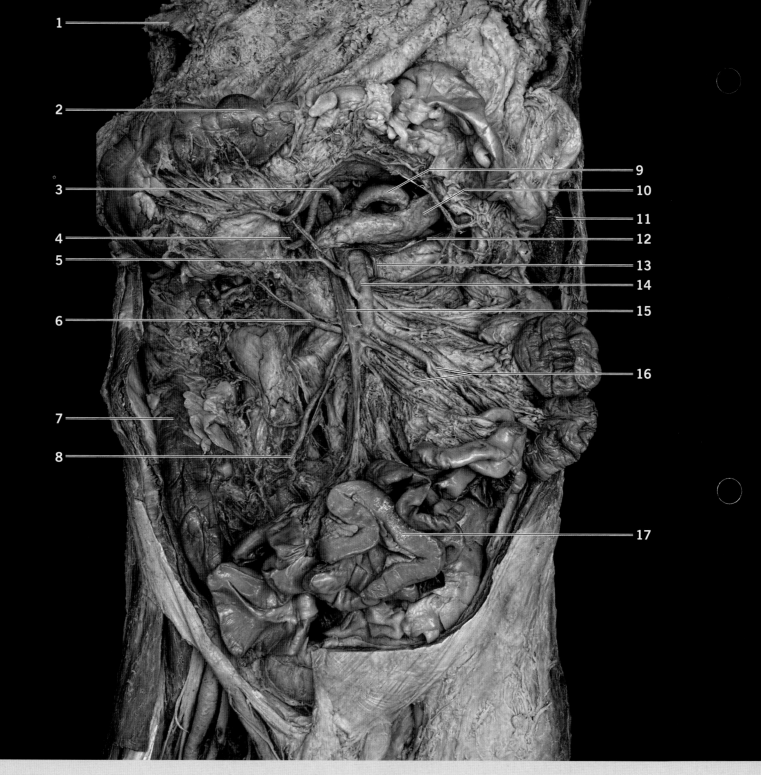

FIGURE 7.10 Major Blood Vessels of the Inferior Abdominal Cavity

1. *Greater omentum*
2. *Transverse colon*
3. Common hepatic artery
4. Gastroduodenal artery
5. Middle colic artery
6. Right colic artery
7. *Ascending colon*
8. Ileocolic artery
9. Splenic artery
10. *Pancreas*
11. *Spleen*
12. Splenic vein
13. Inferior mesenteric vein
14. Superior mesenteric artery
15. Superior mesenteric vein
16. Intestinal arteries
17. *Small intestine*

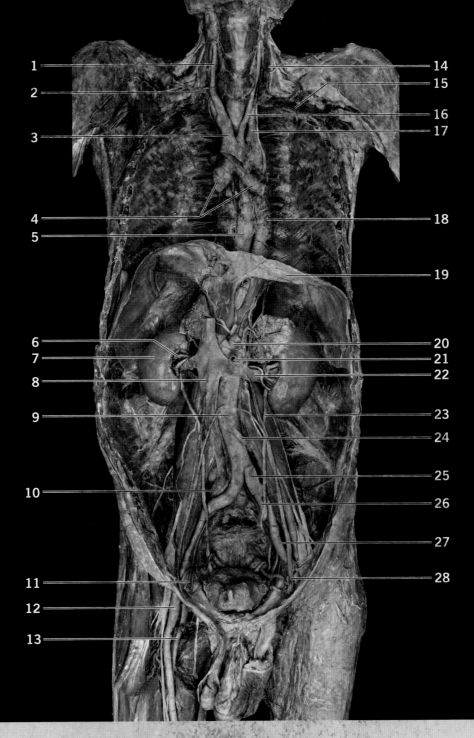

FIGURE 7.11 Major Blood Vessels of the Trunk

1. Right common carotid artery
2. Right subclavian artery
3. Brachiocephalic trunk
4. *Main bronchi*
5. *Esophagus*
6. Renal artery
7. *Kidney*
8. Inferior vena cava
9. Abdominal aorta
10. Common iliac vein
11. External iliac vein
12. Femoral artery
13. Femoral vein
14. Internal jugular vein
15. Subclavian vein
16. Left common carotid artery
17. Left subclavian artery
18. Thoracic aorta
19. *Respiratory diaphragm*
20. Celiac trunk
21. Superior mesenteric artery
22. Renal vein
23. *Ureter*
24. Inferior mesenteric artery
25. Common iliac artery
26. Internal iliac artery
27. External iliac artery
28. Testicular artery and vein

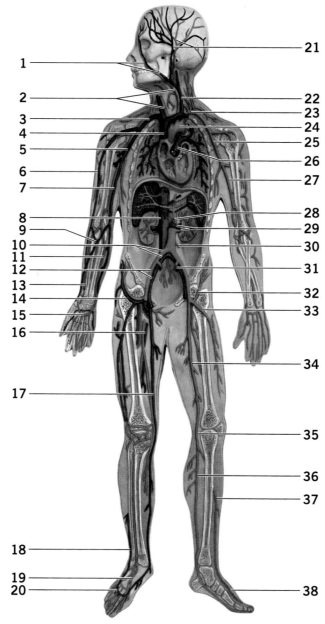

LT-G30: Human Circulatory System, 3B Scientific®

FIGURE 7.12 Major Blood Vessels

1. Facial artery and vein
2. Right & left common carotid arteries
3. Right subclavian vein
4. Superior vena cava
5. Axillary vein
6. Cephalic vein
7. Basilic vein
8. Inferior vena cava
9. Median cubital vein
10. Common iliac artery & vein
11. Ulnar artery
12. External iliac artery & vein
13. Radial artery
14. Superficial circumflex iliac vein
15. Superficial palmar arch
16. Femoral vein
17. Great saphenous vein
18. Small saphenous vein
19. Dorsalis pedis artery
20. Dorsal venous arch
21. Superficial temporal artery & vein
22. Internal jugular vein
23. Left subclavian artery
24. Aorta
25. Axillary artery
26. Pulmonary trunk
27. Brachial artery
28. Superior mesenteric artery
29. Renal artery & vein
30. Inferior mesenteric artery
31. Internal iliac artery & vein
32. Ascending branch of lateral circumflex femoral artery
33. Descending branch of lateral circumflex femoral artery
34. Femoral artery
35. Popliteal artery
36. Posterior tibial artery
37. Anterior tibial artery
38. Dorsalis pedis artery

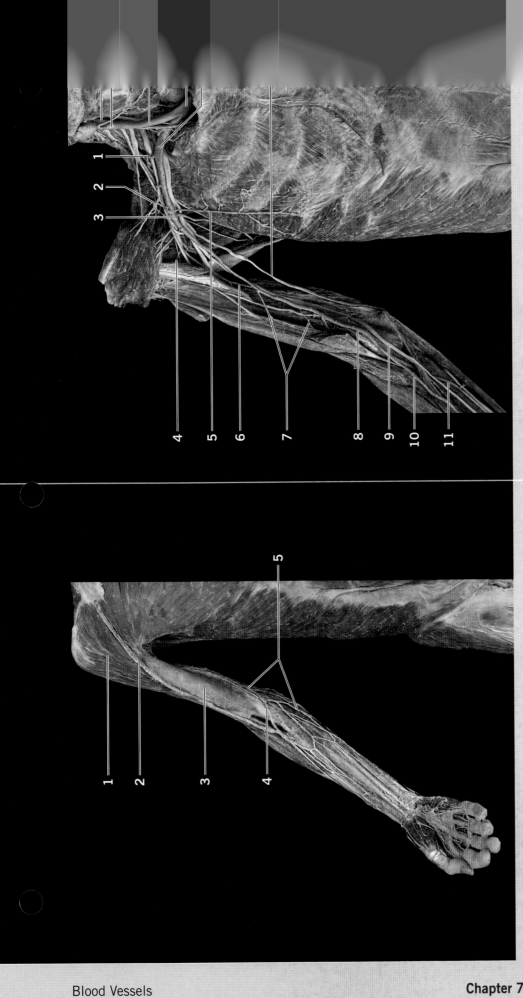

FIGURE 7.14 Major Blood Vessels of the Right Upper Limb

1. Dorsal scapular artery
2. Thoracoacromial trunk
3. Axillary artery
4. Anterior humeral circumflex artery
5. Lateral thoracic artery
6. Deep artery of arm
7. Muscular branches of brachial artery
8. Brachial artery
9. *Median nerve*
10. Radial artery
11. Ulnar artery
12. Internal carotid artery
13. Extern artery
14. Right carotic
15. Brachi trunk
16. Right artery
17. *Ulnar*

FIGURE 7.13 Superficial Veins of the Arm

1. *Deltoid muscle*
2. Cephalic vein
3. *Brachial fascia*
4. Median cubital vein
5. Basilic vein

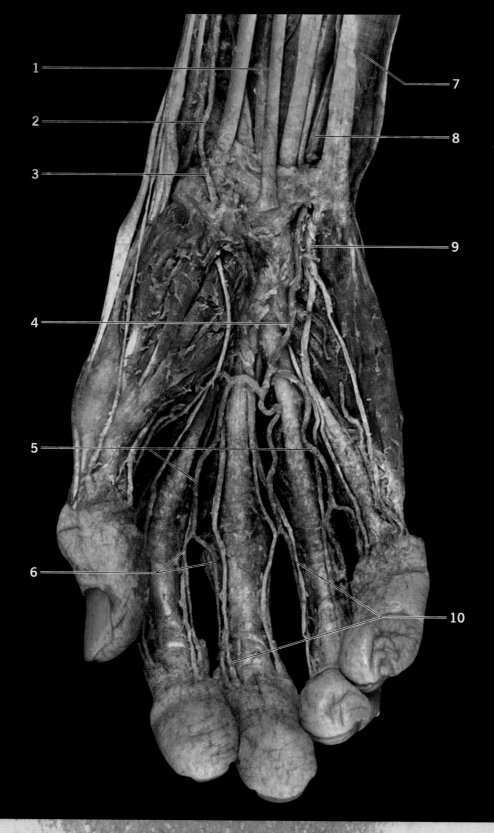

FIGURE 7.15 Superficial Blood Vessels of the Right Hand

1. *Palmaris longus tendon*
2. Radial artery
3. Superficial palmar branch of radial artery
4. Superficial palmar arch
5. Common digital arteries
6. Proper digital arteries
7. *Flexor carpi ulnaris muscle*
8. Ulnar artery
9. *Ulnar nerve*
10. *Proper digital branches of median nerve*

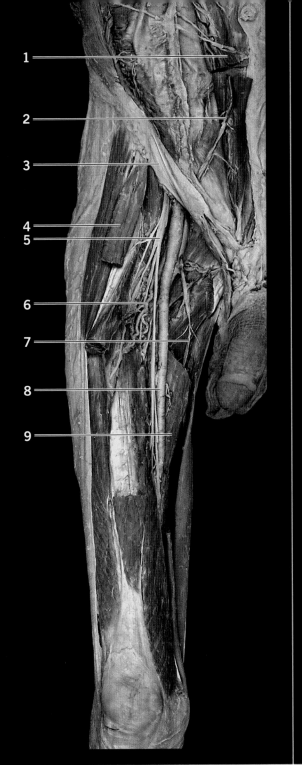

FIGURE 7.16 Blood Vessels of the Right Anterior Thigh

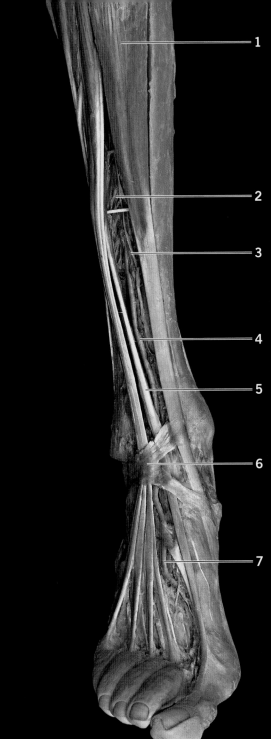

FIGURE 7.17 Blood Vessels of the Right Anterior Lower Leg

1. *Rectus abdominis muscle, cut*
2. Inferior epigastric artery
3. *Inguinal ligament*
4. *Sartorius muscle, cut*
5. *Femoral nerve*
6. Muscular branches of femoral artery
7. Femoral vein tributary
8. Femoral artery
9. *Adductor longus muscle, cut*

1. *Tibialis anterior muscle*
2. *Deep fibular (peroneal) nerve*
3. Anterior tibial vein
4. Anterior tibial artery
5. *Extensor hallucis longus muscle*
6. *Extensor retinaculum*
7. Dorsalis pedis artery

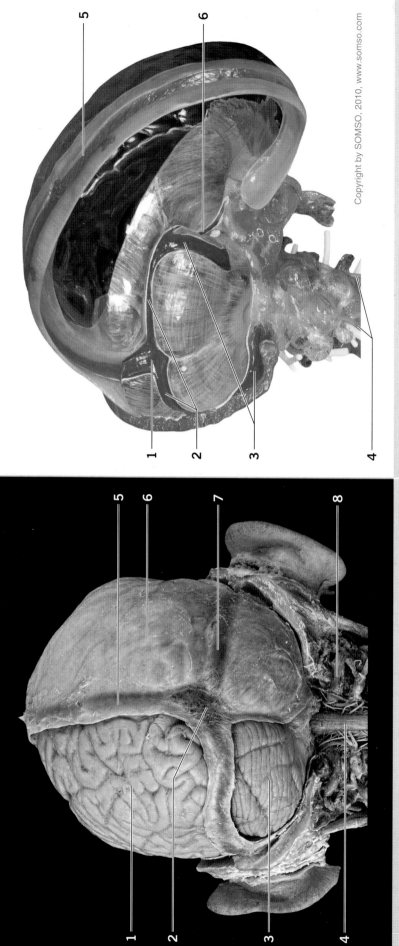

FIGURE 7.19 Dural Sinuses

1. Confluence of the sinuses
2. Transverse sinuses
3. Sigmoid sinuses
4. Vertebral arteries
5. Superior sagittal sinus
6. Superior petrosal sinus

FIGURE 7.18 Dural Sinuses

1. *Left cerebral hemisphere*
2. Confluence of the sinuses
3. *Left cerebellar hemisphere*
4. *Spinal cord*
5. Superior sagittal sinus
6. *Dura mater*
7. Transverse sinus
8. Vertebral artery

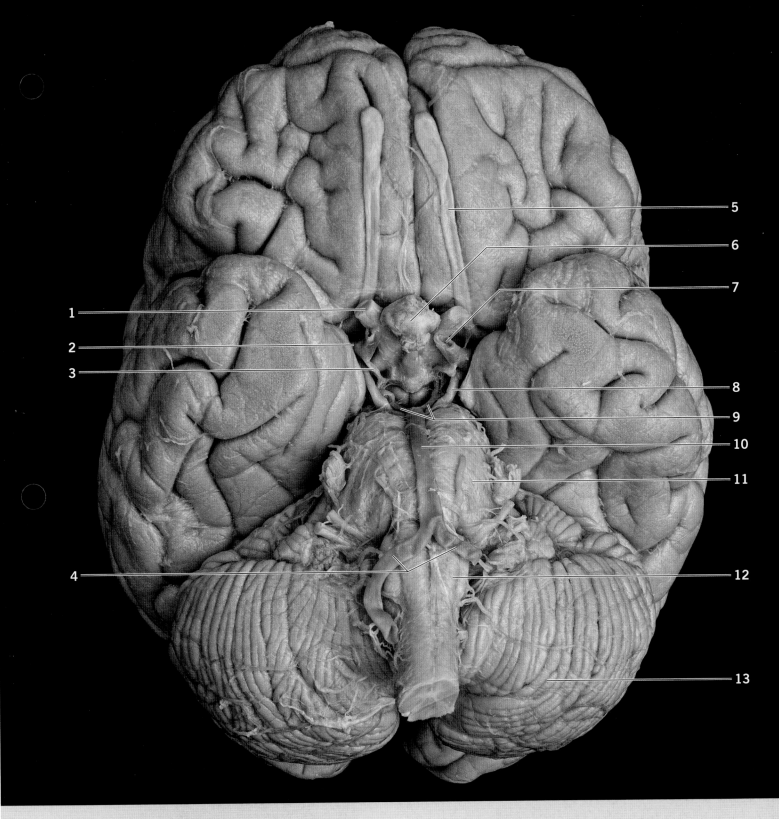

FIGURE 7.20 Circle of Willis

1. *Optic nerve (II)*
2. Middle cerebral artery
3. Posterior communicating artery
4. Vertebral arteries
5. *Olfactory tract*
6. *Pituitary gland*
7. Internal carotid artery
8. *Oculomotor nerve (III)*
9. Posterior cerebral arteries
10. Basilar artery
11. *Pons*
12. *Medulla oblongata*
13. *Cerebellum*

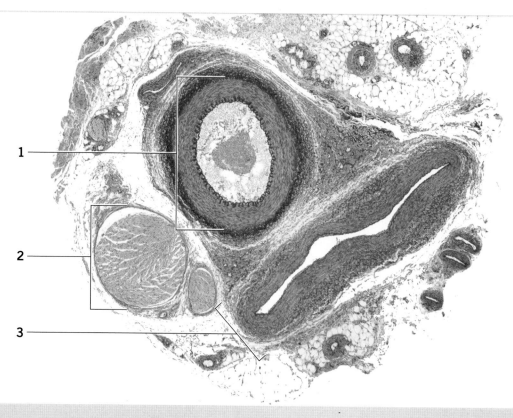

FIGURE 7.21 Artery, Vein, and Nerve, Cross Section, 40×

1. Artery
2. *Nerve*
3. Vein

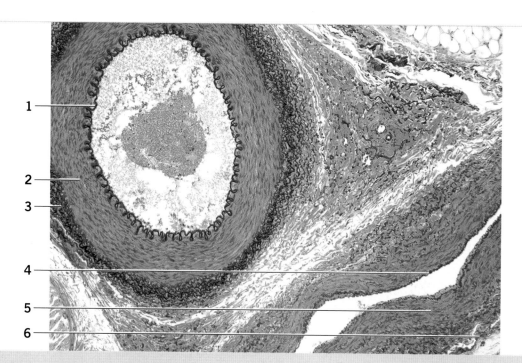

FIGURE 7.22 Artery and Vein, Cross Section, 100×

1. Tunica intima of artery
2. Tunica media of artery
3. Tunica externa of artery
4. Tunica intima of vein
5. Tunica media of vein
6. Tunica externa of vein

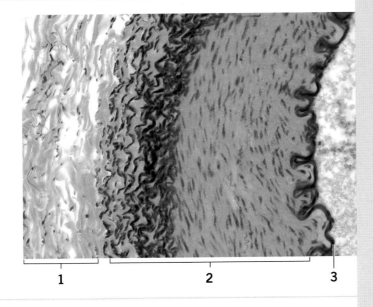

FIGURE 7.23 Artery, Cross Section, 400×

1. Tunica externa of artery
2. Tunica media of artery
3. Tunica intima of artery

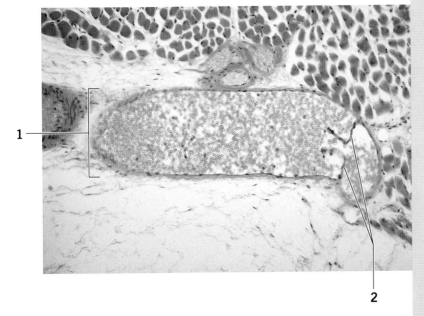

FIGURE 7.24 Vein Showing Valve, Longitudinal Section, 200×

1. Venule
2. Venous valves

FIGURE 7.25 Cardiac Muscle, Longitudinal Section, 1000×

1. Cardiac muscle cell
2. Nucleus
3. Intercalated discs

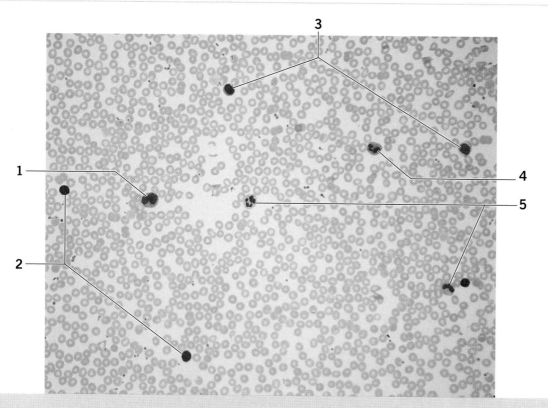

FIGURE 7.26 Blood Smear, 400×

1. Monocyte

2. Basophils

3. Lymphocytes

4. Eosinophil

5. Neutrophils

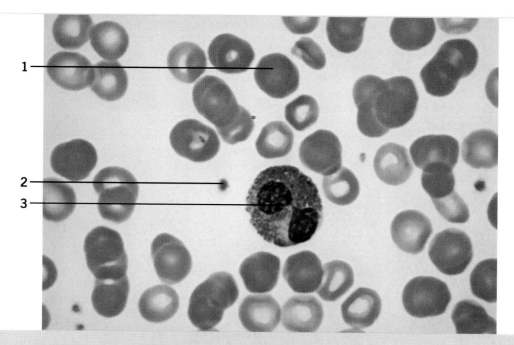

FIGURE 7.27 Blood Smear, 1200×

1. Red blood cell

2. Platelet

3. Eosinophil

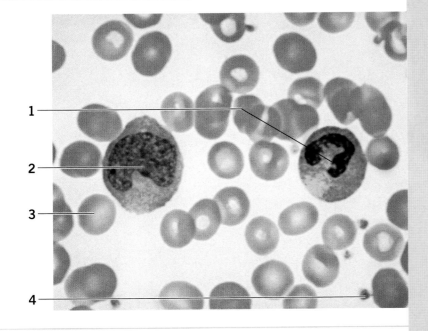

FIGURE 7.28 Blood Smear, 1200×

1. Neutrophil
2. Monocyte
3. Red blood cell
4. Platelet

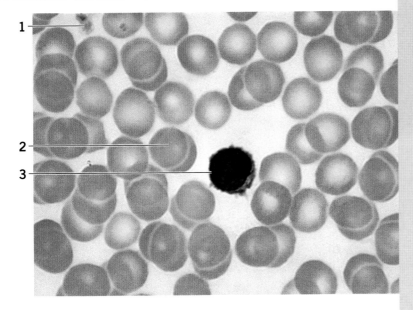

FIGURE 7.29 Blood Smear, 1200×

1. Platelet
2. Red blood cell
3. Lymphocyte

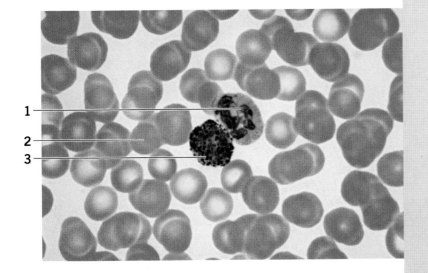

FIGURE 7.30 Blood Smear, 1200×

1. Neutrophil
2. Red blood cell
3. Basophil

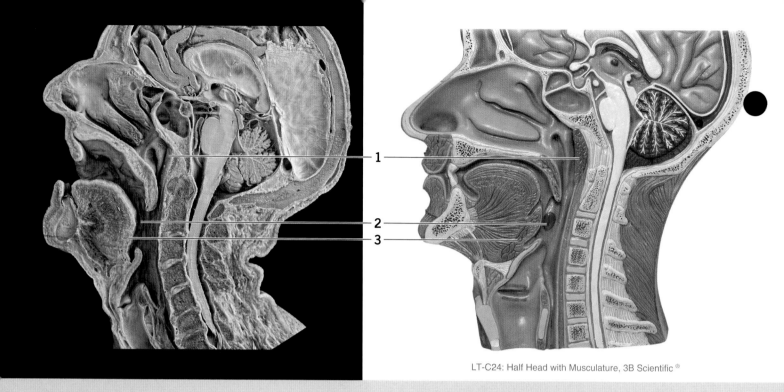

LT-C24: Half Head with Musculature, 3B Scientific ®

FIGURE 8.1 Head, Midsagittal View

1. Pharyngeal tonsil   **2.** Palatine tonsil   **3.** Lingual tonsil

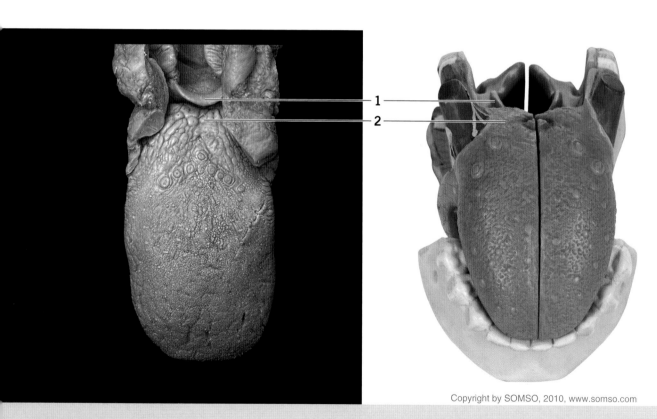

Copyright by SOMSO, 2010, www.somso.com

FIGURE 8.2 Tongue, Superior View

1. *Epiglottis*   **2.** Lingual tonsil

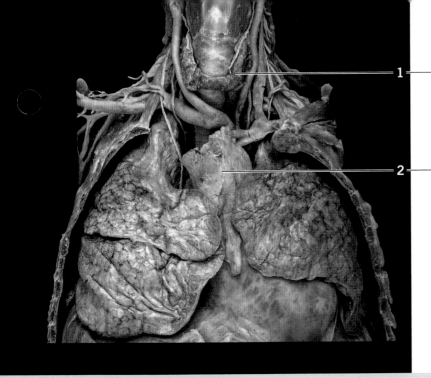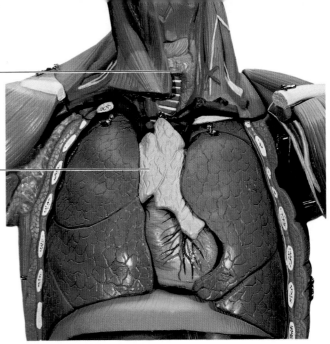

FIGURE 8.3 Thymus

1. *Thyroid gland*

2. Thymus

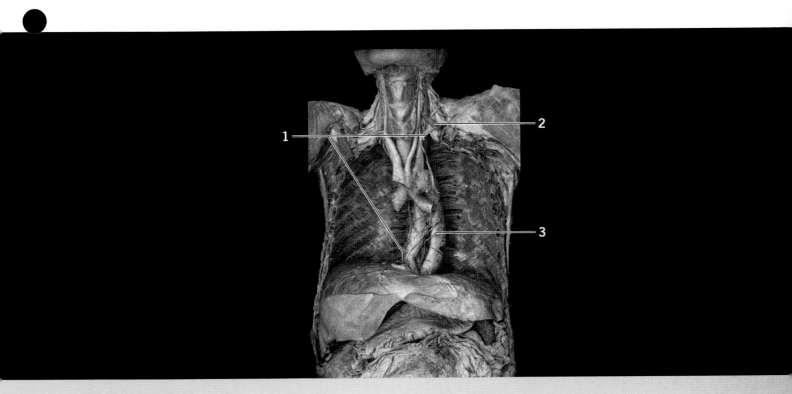

FIGURE 8.4 Thoracic Duct

1. Thoracic duct

2. *Left subclavian vein*

3. Thoracic aorta

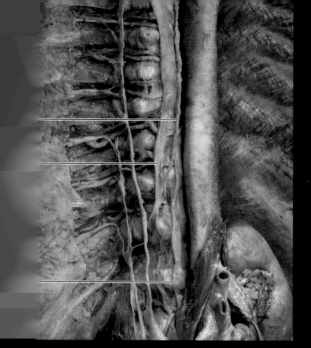

FIGURE 8.5 Thoracic Duct and Cisterna Chyli

1. Thoracic duct
2. *Azygos vein*
3. Cisterna chyli

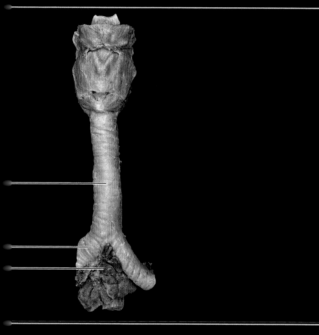

FIGURE 8.6 Tracheobronchial Lymph Nodes

1. *Trachea*
2. *Right primary bronchus*
3. Tracheobronchial lymph nodes

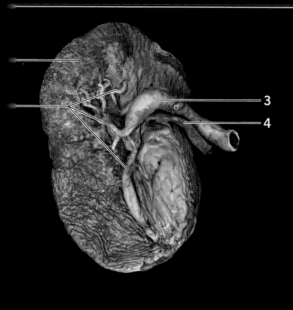

FIGURE 8.7 Spleen, Medial View

1. Spleen
2. *Branches of splenic artery*
3. *Splenic artery*
4. *Splenic vein*

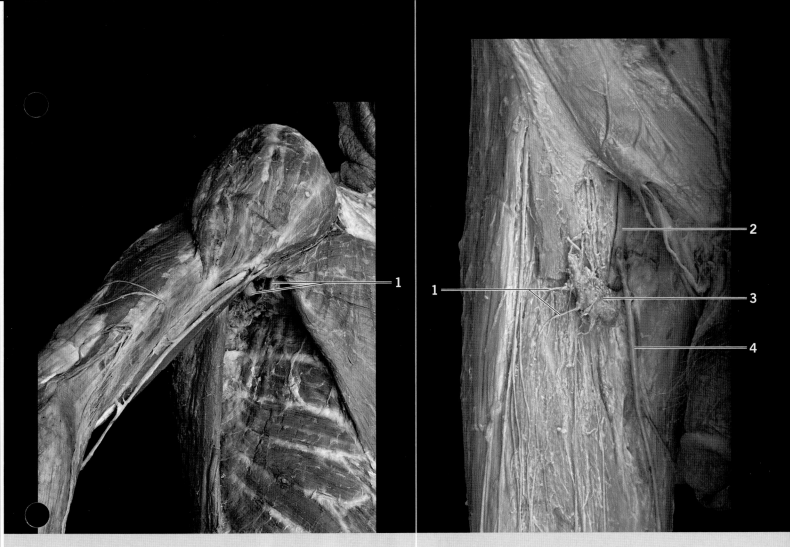

FIGURE 8.8 Axillary Lymph Nodes

FIGURE 8.9 Inguinal Lymph Node and Lymphatic Collecting Vessels

1. Axillary lymph nodes

1. Lymphatic collecting vessels
2. *Femoral vein*

3. Superficial inguinal lymph node
4. *Great saphenous vein*

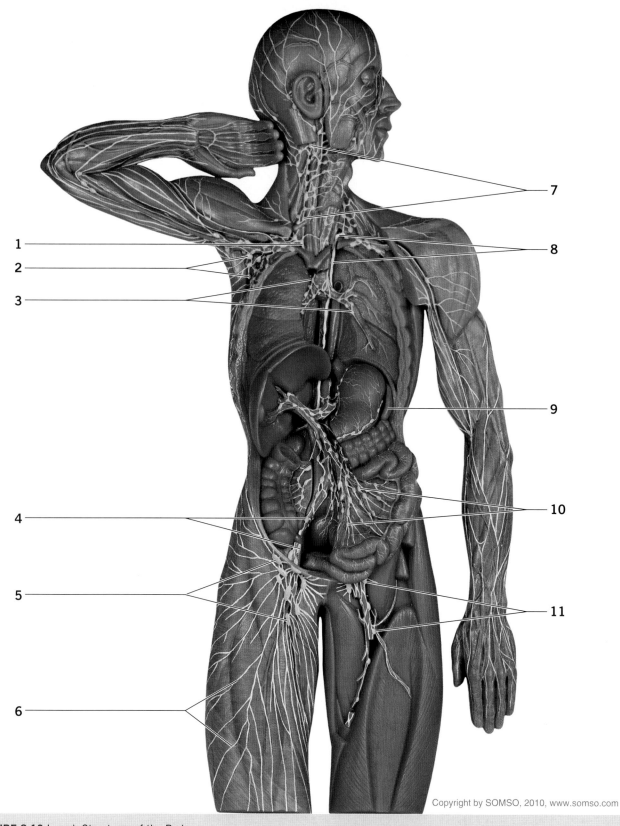

Copyright by SOMSO, 2010, www.somso.com

FIGURE 8.10 Lymph Structures of the Body

1. Right lymphatic duct
2. Axillary lymph nodes
3. Mediastinal lymph nodes
4. Common iliac lymph nodes
5. Superficial inguinal lymph nodes
6. Lymphatic collecting vessels
7. Deep cervical lymph nodes
8. Thoracic duct
9. Spleen
10. Superior mesenteric lymph nodes
11. Deep inguinal lymph nodes

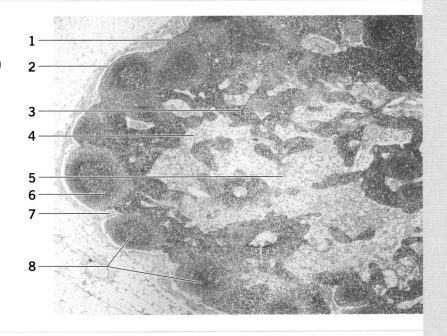

FIGURE 8.11 Lymph Node, Cross Section, 40×

1. Cortex
2. Capsule
3. Medullary cord
4. Medullary sinus
5. Medulla
6. Germinal center
7. Trabecula
8. Lymphoid follicles

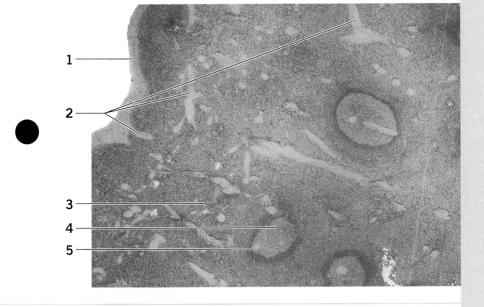

FIGURE 8.12 Spleen, Cross Section, 40×

1. Capsule
2. Trabecular vein
3. Red pulp (splenic cords)
4. White pulp (periarteriolar lymphocyte sheath)
5. Central artery

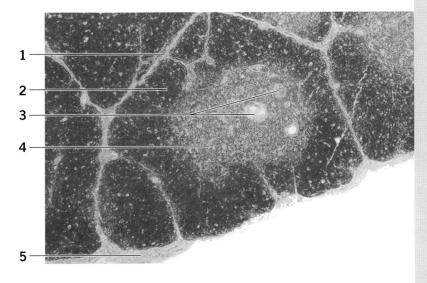

FIGURE 8.13 Thymus, Cross Section, 100×

1. Thymic septum
2. Cortex
3. Thymic corpuscles
4. Medulla
5. Capsule

Tissues of the Lymphatic System

Chapter 8 The Lymphatic System   **157**

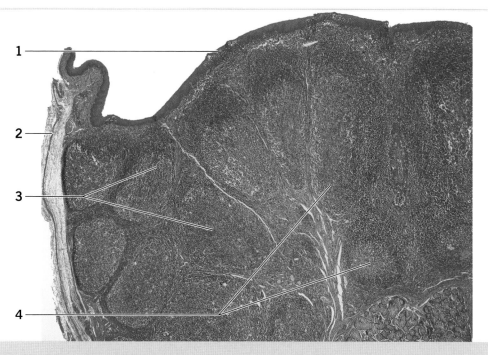

FIGURE 8.14 Palatine Tonsil, Cross Section, 40×

1. Stratified squamous epithelium
2. Capsule
3. Lymphoid follicles
4. Germinal centers

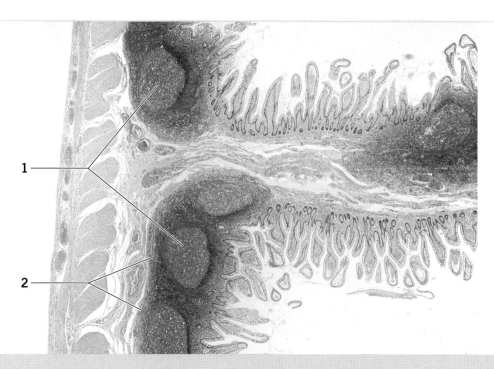

FIGURE 8.15 Peyer's Patches, Ileum, Cross Section, 40×

1. Germinal centers
2. Peyer's patches

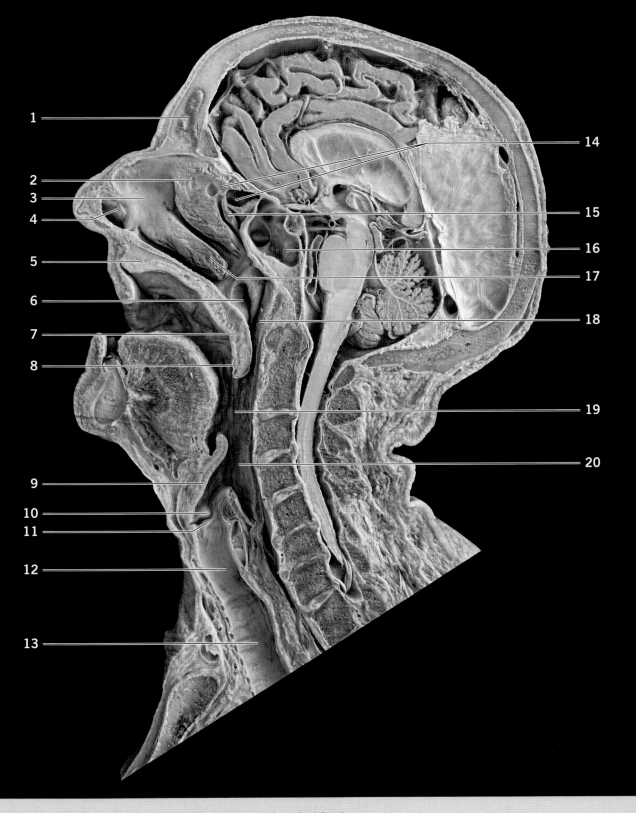

FIGURE 9.1 Respiratory Structures of the Head and Neck, Midsagittal Section

1. Frontal sinus
2. Middle nasal concha
3. Nasal cavity
4. Nasal vestibule
5. Hard palate
6. Opening of pharyngotympanic tube
7. Soft palate
8. Uvula
9. Epiglottis
10. Vestibular fold
11. Vocal fold
12. Larynx
13. Trachea
14. Ethmoid air cells
15. Superior nasal concha
16. Sphenoid sinus
17. Inferior nasal concha
18. Nasopharynx
19. Oropharynx
20. Laryngopharynx

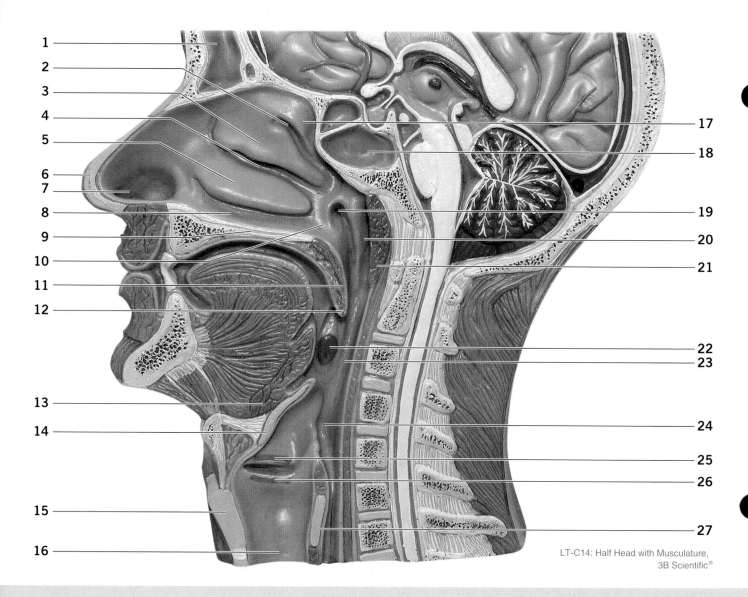

1
2
3
4
5
6
7
8
9
10
11
12
13
14
15
16
17
18
19
20
21
22
23
24
25
26
27

LT-C14: Half Head with Musculature,
3B Scientific®

FIGURE 9.2 Respiratory Structures of the Head and Neck, Midsagittal Section

1. Frontal sinus
2. Superior meatus
3. Middle concha
4. Middle meatus
5. Inferior concha
6. Cartilage of nose
7. Nasal vestibule
8. Inferior meatus
9. Hard palate
10. Internal naris
11. Soft palate
12. Uvula
13. Lingual tonsil
14. Epiglottis
15. Thyroid cartilage
16. Larynx
17. Superior concha
18. Sphenoidal sinus
19. Opening of
    pharyngotympanic tube
20. Nasopharynx
21. Pharyngeal tonsil
22. Palatine tonsil
23. Oropharynx
24. Laryngopharynx
25. Vestibular fold
26. Vocal fold
27. Cricoid cartilage

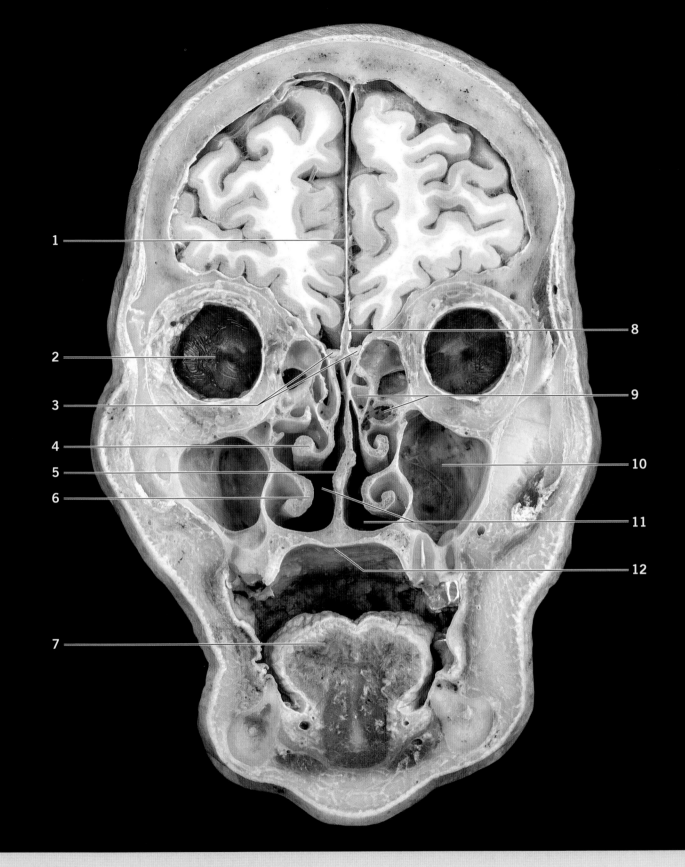

FIGURE 9.3 Respiratory Structures of the Head, Frontal Section

1. *Falx cerebri*
2. *Posterior cavity of eye*
3. Olfactory bulbs
4. Middle nasal concha
5. Nasal septum
6. Inferior nasal concha
7. *Tongue*
8. *Crista galli*
9. Ethmoid air cells
10. Maxillary sinus
11. Nasal cavity
12. Hard palate

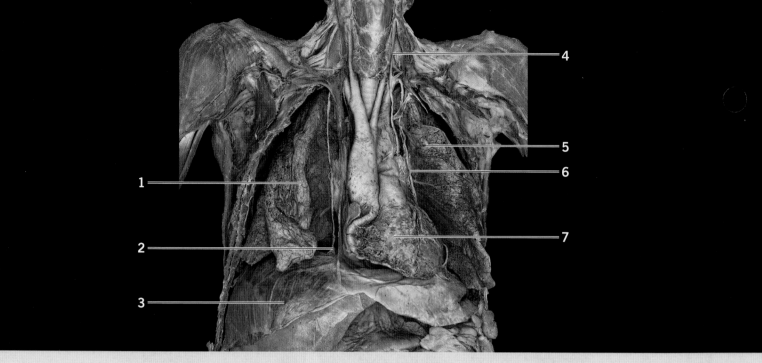

FIGURE 9.4 Respiratory Structures of the Thoracic Cavity

1. Right lung
2. *Right phrenic nerve*
3. Respiratory diaphragm
4. *Left vagus nerve*
5. Left lung
6. *Left phrenic nerve*
7. *Heart*

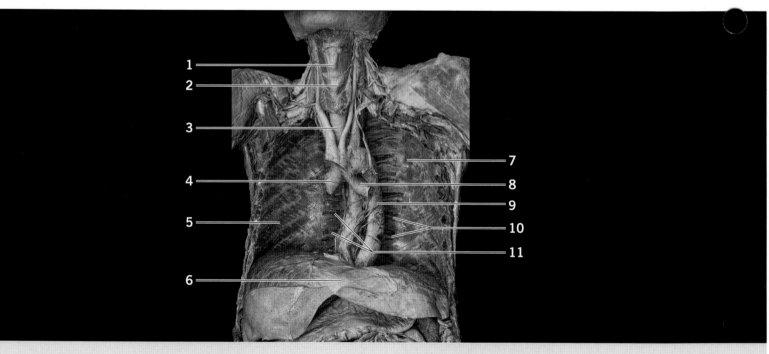

FIGURE 9.5 Thoracic Cavity with Lungs and Heart Removed

1. Thyroid cartilage
2. Cricoid cartilage
3. Trachea
4. Right main bronchus
5. Innermost intercostal muscle
6. Respiratory diaphragm
7. Subcostal muscles
8. Left main bronchus
9. *Aorta*
10. *Intercostal nerves*
11. *Posterior intercostal arteries and veins*

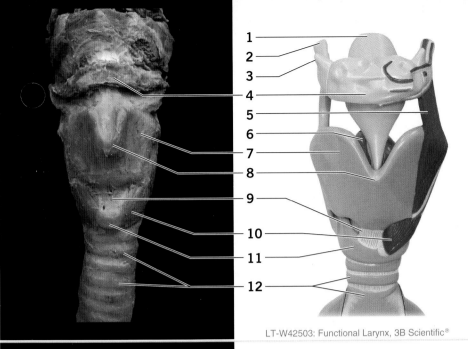

LT-W42503: Functional Larynx, 3B Scientific®

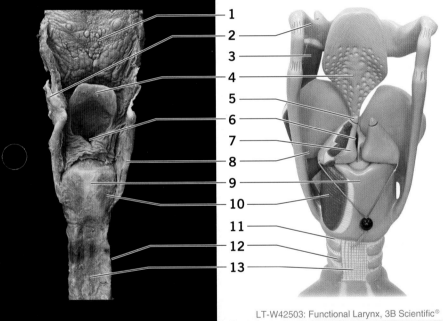

LT-W42503: Functional Larynx, 3B Scientific®

FIGURE 9.7 Larynx, Posterior View

1. *Tongue*
2. Hyoid bone
3. Thyrohyoid muscle
4. Epiglottis
5. Corniculate cartilage
6. Rima glottidis
7. Arytenoid cartilage
8. Thyroid cartilage
9. Cricoid cartilage
10. Posterior cricoarytenoid muscle
11. Cricotracheal ligament
12. Tracheal cartilage
13. Muscular part of trachea

FIGURE 9.8 Larynx, Internal View from Posterior

1. Epiglottis
2. Thyroid cartilage
3. Cricoid cartilage
4. Trachea
5. *Tongue*
6. Vestibular fold
7. Vocal fold

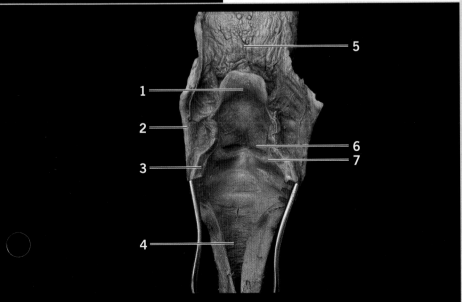

FIGURE 9.9 Right and Left Lungs, Anterior View

1. Apex of right lung
2. Superior lobe of right lung
3. Horizontal fissure of right lung
4. Middle lobe of right lung
5. Inferior lobe of right lung
6. Oblique fissure of right lung
7. Apex of left lung
8. Superior lobe of left lung
9. Inferior lobe of left lung
10. Oblique fissure of left lung
11. Cardiac notch

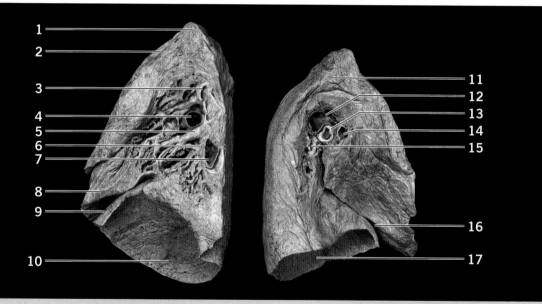

FIGURE 9.10 Right and Left Lungs, Medial Views

1. Apex of right lung
2. Superior lobe of right lung
3. Right superior lobarbronchus
4. Right pulmonary artery
5. Right middle lobar bronchus
6. Right inferior lobar bronchus
7. Right inferior pulmonary vein
8. Middle lobe
9. Oblique fissure of right lung
10. Base of right lung, diaphragmatic surface
11. Superior lobe of left lung
12. Left pulmonary artery
13. Left main bronchus
14. Left superior pulmonary vein
15. Left inferior lobar bronchus
16. Oblique fissure of left lung
17. Base of left lung, diaphragmatic surface

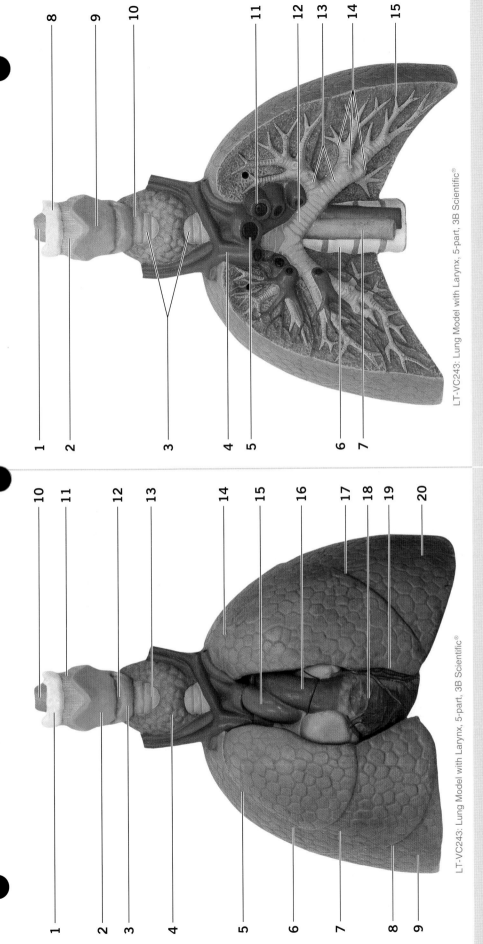

LT-VC243: Lung Model with Larynx, 5-part, 3B Scientific®

FIGURE 9.12 Deep View of Lungs Showing Branching of Bronchi

1. Epiglottis
2. Thyrohyoid membrane
3. Trachea
4. *Superior vena cava*
5. *Aorta*
6. *Vertebra*
7. *Esophagus*
8. Hyoid bone
9. Larynx
10. Cricotracheal ligament
11. *Pulmonary artery*
12. Main bronchus
13. Lobar bronchi
14. Segmental bronchi
15. Higher-order bronchus

LT-VC243: Lung Model with Larynx, 5-part, 3B Scientific®

FIGURE 9.11 Larynx, Trachea, and Lungs

1. Hyoid bone
2. Thyroid cartilage
3. Cricoid cartilage
4. *Thyroid gland*
5. Superior lobe of right lung
6. Horizontal fissure of right lung
7. Middle lobe of right lung
8. Oblique fissure of right lung
9. Inferior lobe of right lung
10. Epiglottis of larynx
11. Thyrohyoid membrane
12. Cricothyroid ligament
13. Trachea
14. Superior lobe of left lung
15. *Aorta*
16. *Pulmonary trunk*
17. Oblique fissure of left lung
18. *Heart*
19. Cardiac notch of left lung
20. Inferior lobe of left lung

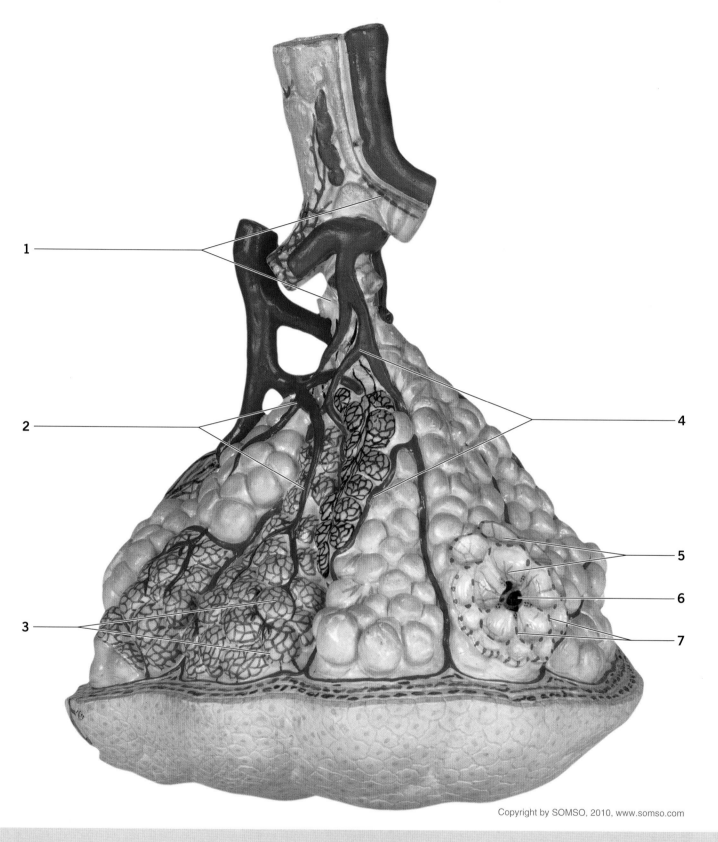

1

2

3

4

5

6

7

FIGURE 9.13 Alveolar Sac

1. Bronchioles
2. Pulmonary arterioles
3. Alveoli surrounded by capillaries
4. Pulmonary venules
5. Alveoli
6. Opening of alveolar duct
7. Pulmonary capillaries

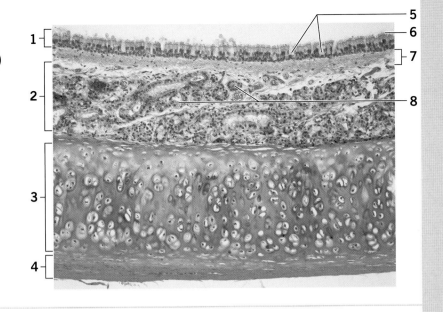

FIGURE 9.14 Cross Section of Trachea, 100×

1. Pseudostratified columnar epithelium
2. Submucosa
3. Hyaline cartilage
4. Perichondrium
5. Goblet cells
6. Cilia
7. Lamina propria
8. Seromucous glands

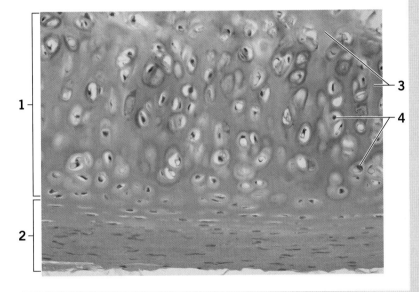

FIGURE 9.15 Cross Section of Trachea, 400×

1. Hyaline cartilage
2. Perichondrium
3. Matrix
4. Chondrocytes in lacunae

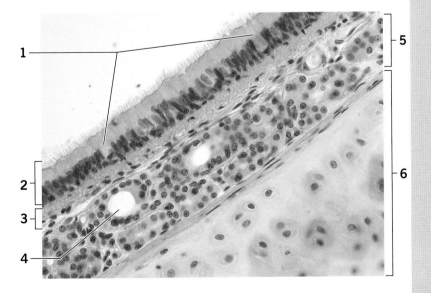

FIGURE 9.16 Cross Section of Trachea, 400×

1. Goblet cells
2. Pseudostratified columnar epithelium with cilia and goblet cells
3. Lamina propria
4. Seromucous gland
5. Submucosa
6. Hyaline cartilage

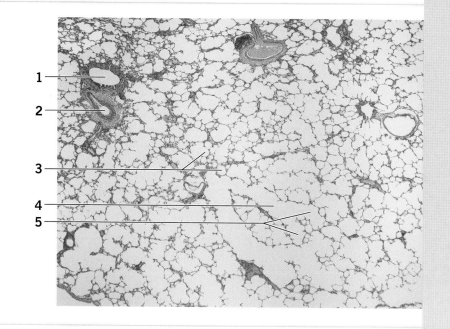

FIGURE 9.17 Cross Section of Lung, 40×

1. Bronchiole
2. Arterial blood vessel
3. Alveoli
4. Alveolar duct
5. Alveolar sacs

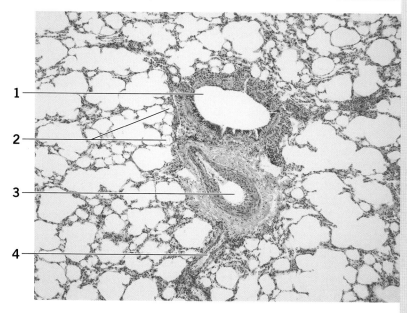

FIGURE 9.18 Cross Section of Lung, 100×

1. Bronchiole
2. Alveoli
3. Arterial blood vessel
4. Venous blood vessel

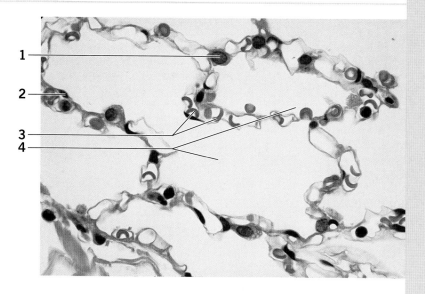

FIGURE 9.19 Alveoli of Lung, 1000×

1. Type II cell
2. Type I cell
3. Red blood cells in capillary
4. Alveolar lumina

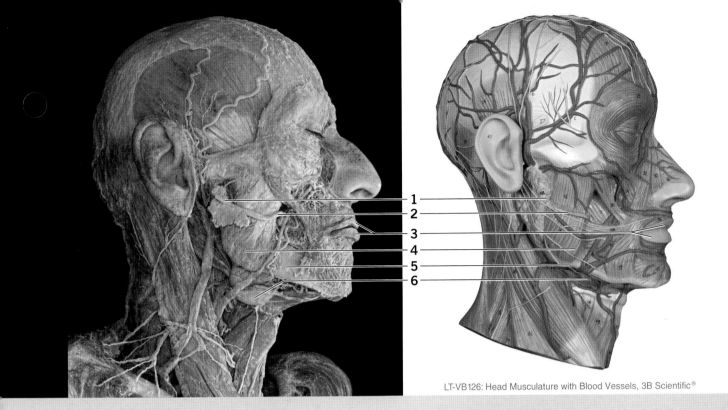

FIGURE 10.1 Head and Neck, Lateral View

LT-VB126: Head Musculature with Blood Vessels, 3B Scientific®

1. Parotid gland
2. Parotid duct
3. Labia
4. Masseter muscle
5. Mandible
6. Submandibular gland

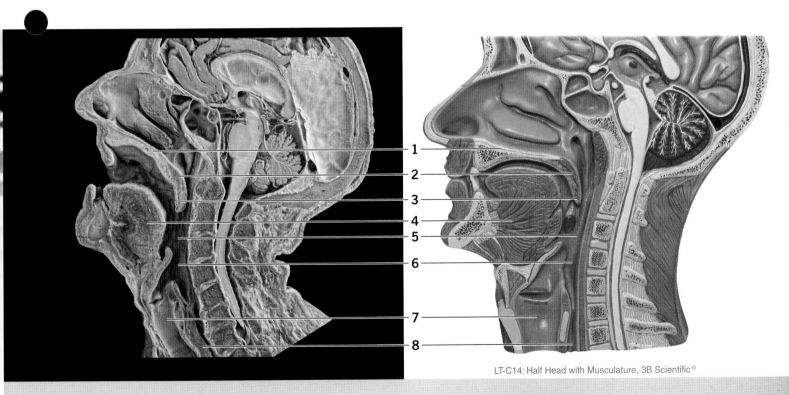

FIGURE 10.2 Head and Neck, Midsagittal View

LT-C14: Half Head with Musculature, 3B Scientific®

1. Hard palate
2. Soft palate
3. Uvula
4. Tongue
5. Oropharynx
6. Laryngopharynx
7. Larynx
8. Esophagus

Gross Anatomy of the Digestive System

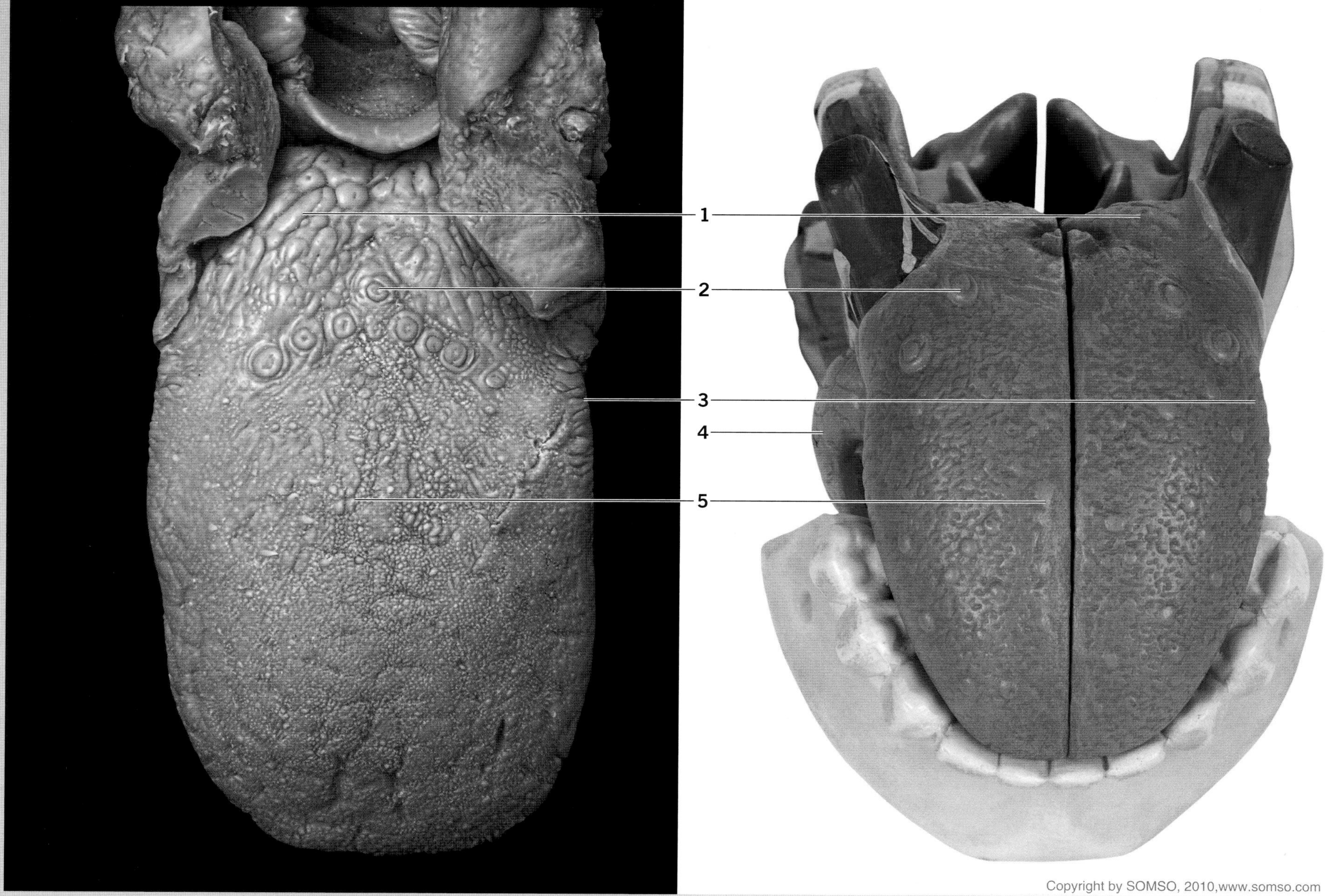

FIGURE 10.3 Tongue, Superior View

1. Lingual tonsil
2. Vallate papilla
3. Foliate papilla
4. Submandibular gland
5. Fungiform papilla

Gross Anatomy of the Digestive System

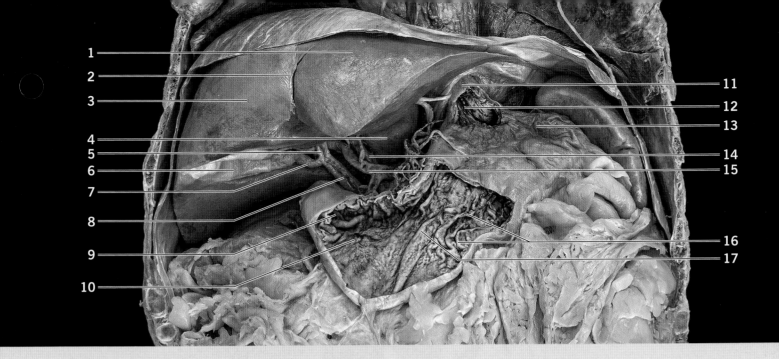

FIGURE 10.4 Upper Abdominal Cavity

1. Left lobe of liver	**6.** Gallbladder	**11.** Esophagus	**16.** Rugae
2. Falciform ligament	**7.** Cystic duct	**12.** Cardiac orifice	**17.** Body of stomach
3. Right lobe of liver	**8.** Bile duct	**13.** Fundus of stomach	
4. Quadrate lobe of liver	**9.** Pyloric orifice	**14.** Hepatic portal vein	
5. Common hepatic duct	**10.** Pylorus of stomach	**15.** Hepatic artery proper	

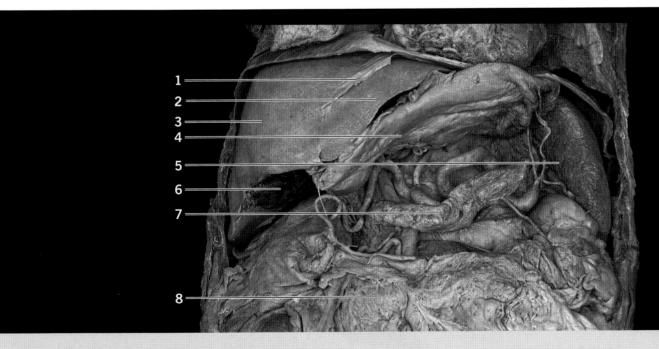

FIGURE 10.5 Upper Abdominal Cavity

1. Falciform ligament	**3.** Right lobe of liver	**5.** *Spleen*	**7.** Pancreas
2. Left lobe of liver	**4.** Stomach	**6.** Gallbladder	**8.** Greater omentum

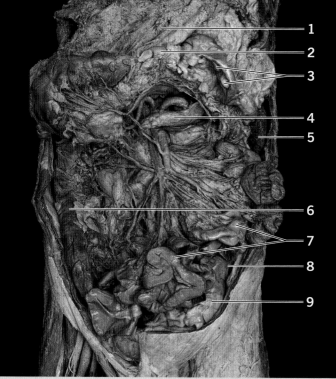

FIGURE 10.6 Abdominal Cavity

1. Greater omentum
2. Transverse colon
3. Epiploic appendages
4. Pancreas
5. *Spleen*
6. Ascending colon
7. Small intestine
8. Descending colon
9. Sigmoid colon

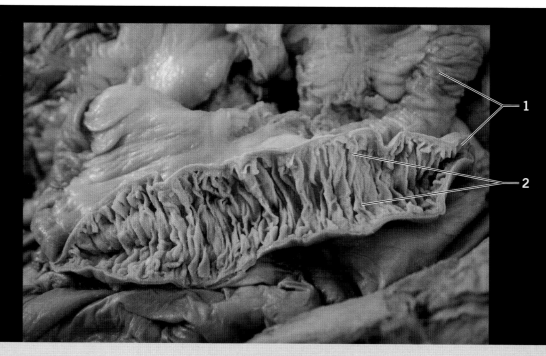

FIGURE 10.7 Small Intestine, Opened

1. Small intestine
2. Plicae circulares/circular folds

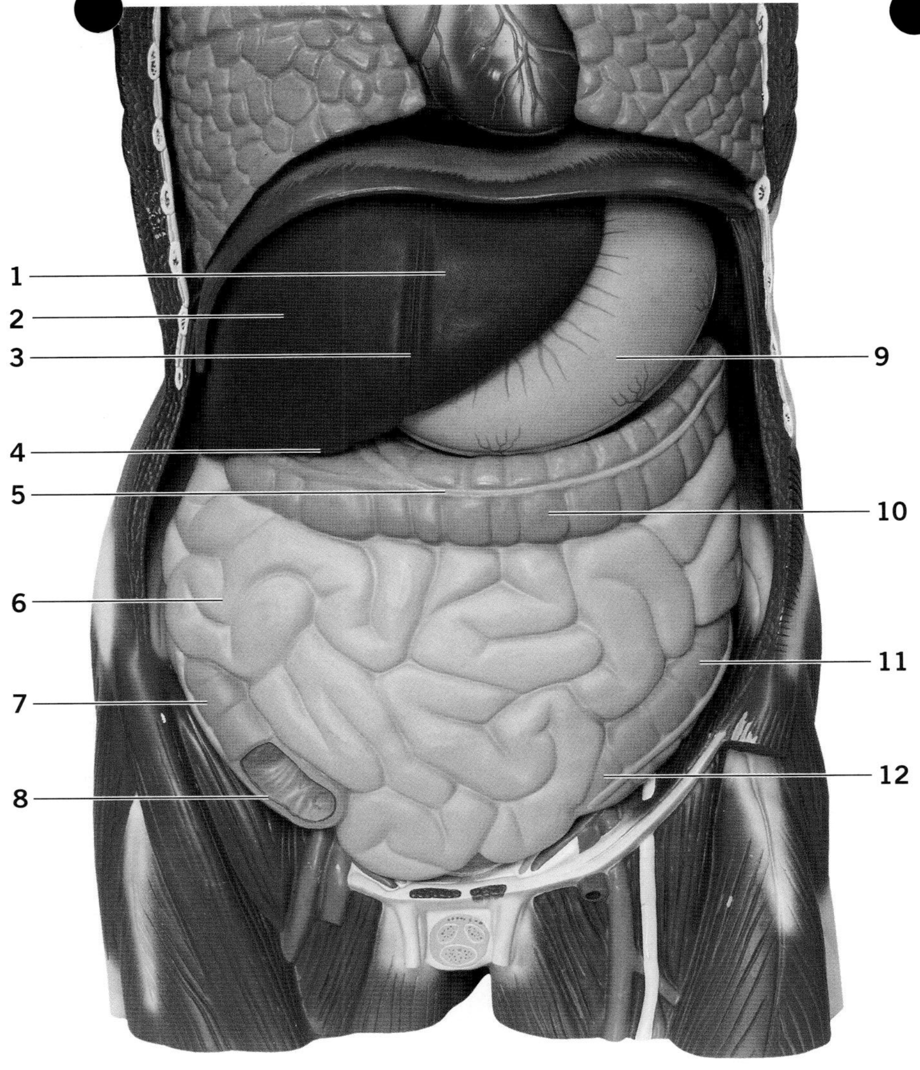

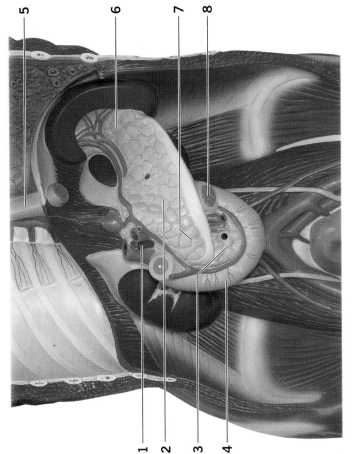

LT-VA16: Life-size Muscle Torso, 27-part, 3B Scientific®

FIGURE 10.9 Upper Abdominal Cavity, Stomach and Lever Removed

1. Bile duct
2. Body of pancreas
3. Head of pancreas
4. Duodenum
5. Esophagus
6. Tail of pancreas
7. Neck of pancreas
8. Jejunum

LT-VA16: Life-size Muscle Torso, 27-part, 3B Scientific®

FIGURE 10.8 Abdominal Cavity

1. Left lobe of liver
2. Right lobe of liver
3. Falciform ligament
4. Gallbladder
5. Teniae coli
6. Small intestine
7. Ascending colon
8. Cecum
9. Stomach
10. Transverse colon
11. Descending colon
12. Sigmoid colon

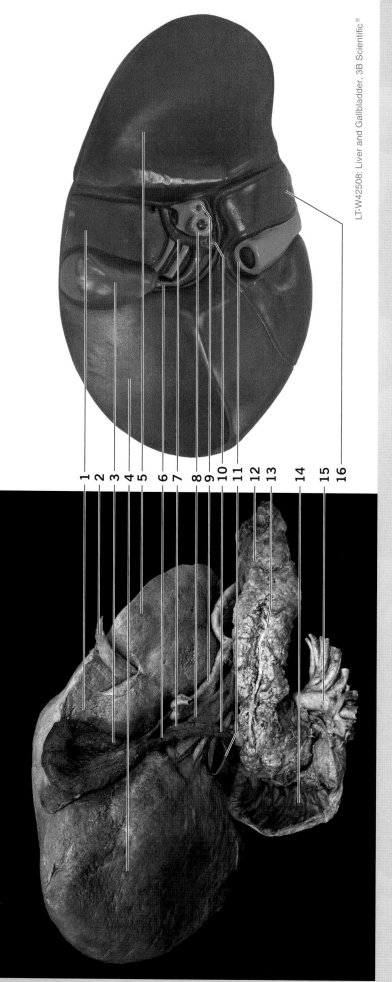

LT-W42508: Liver and Gallbladder, 3B Scientific®

FIGURE 10.10 Liver and Gallbladder, Inferior View; Duodenum and Pancreas, Anterior View; and Liver and Gallbladder, Inferior View

1. Quadrate lobe of liver
2. Ligamentum teres
3. Gallbladder
4. Right lobe of liver

5. Left lobe of liver
6. Cystic duct
7. Common hepatic duct
8. Hepatic portal vein

9. Common hepatic artery
10. Bile duct
11. Inferior vena cava
12. Pancreas

13. Main pancreatic duct
14. Duodenum
15. Superior mesenteric artery
16. Caudate lobe of liver

Gross Anatomy of the Digestive System

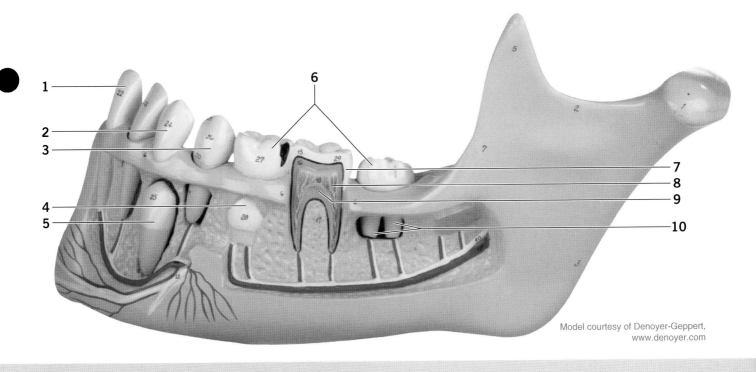

Model courtesy of Denoyer-Geppert,
www.denoyer.com

FIGURE 10.11 Lower Jaw Dentition

1. Incisor
2. Deciduous canine
3. Premolar

4. Unerupted permanent premolar
5. Unerupted permanent canine

6. Molars
7. Enamel
8. Pulp

9. Dentin
10. Roots of molar

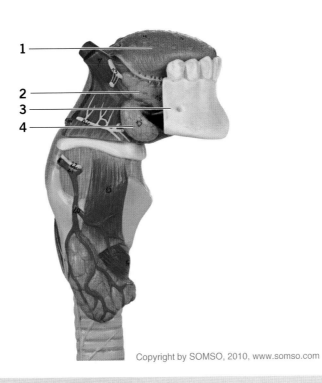

Copyright by SOMSO, 2010, www.somso.com

FIGURE 10.12 Salivary Glands

1. Tongue

2. Sublingual gland

3. Mandible

4. Submandibular gland

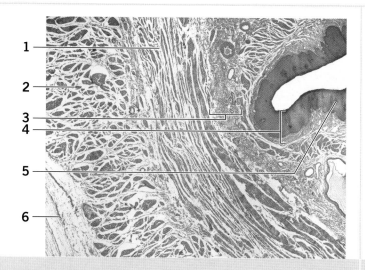

FIGURE 10.13 Esophagus, Lower Third, Cross Section, 40×

1. Circular muscle of muscularis externa
2. Longitudinal muscle of muscularis externa
3. Submucosa
4. Mucosa
5. Stratified squamous epithelium
6. Adventitia

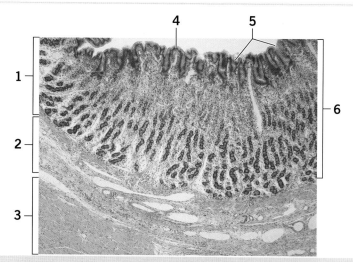

FIGURE 10.14 Fundic Stomach, Cross Section, 100×

1. Mucosa
2. Submucosa
3. Oblique layer of muscularis externa
4. Simple columnar epithelium
5. Gastric pits
6. Gastric glands in lamina propria

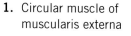
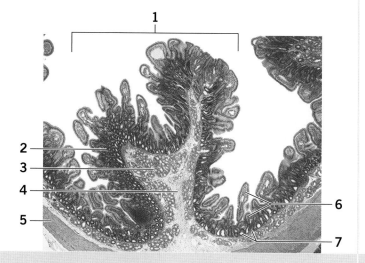

FIGURE 10.15 Duodenum, Cross Section, 40×

1. Plica circulares/circular fold
2. Mucosa of plica circulares
3. Duodenal glands
4. Submucosa
5. Muscularis externa
6. Villi
7. Intestinal crypts

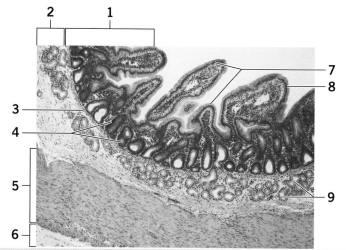

FIGURE 10.16 Duodenum, Cross Section, 100×

1. Mucosa
2. Submucosa
3. Muscularis mucosae
4. Intestinal crypts
5. Muscularis externa
6. Serosa or adventitia
7. Villi
8. Simple columnar epithelium
9. Duodenal glands

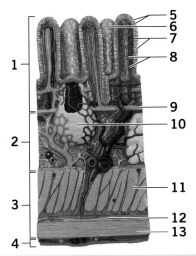

FIGURE 10.17 Wall of Small Intestine, Ileum

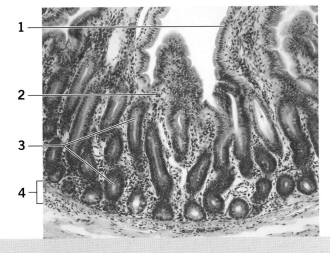

FIGURE 10.18 Jejunum, Cross Section, 200×

1. Mucosa
2. Submucosa
3. Muscularis externa
4. Serosa or adventitia
5. Goblet cells
6. Villus
7. Simple columnar epithelium
8. Capillaries
9. Muscularis mucosae
10. Peyer's patch
11. Circular muscle of muscularis externa
12. Myenteric plexus
13. Longitudinal muscle of muscularis externa

1. Simple columnar epithelium
2. Lamina propria
3. Intestinal crypts
4. Muscularis mucosae

Model courtesy of Ward's Natural Science Establishment, Inc. www.wardsci.com

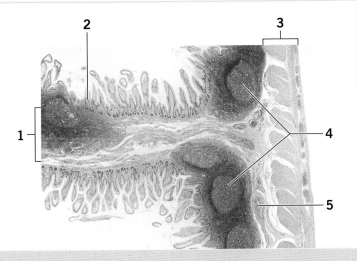

FIGURE 10.19 Ileum, Cross Section, 40×

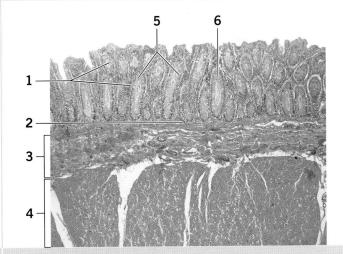

FIGURE 10.20 Large Intestine, Cross Section, 100×

1. Submucosa of plica circulares/circular fold
2. Plica circulares
3. Muscularis externa
4. Peyer's patches
5. Submucosa

1. Goblet cells
2. Muscularis mucosae
3. Submucosa
4. Circular muscle of muscularis externa
5. Intestinal crypts
6. Simple columnar epithelium

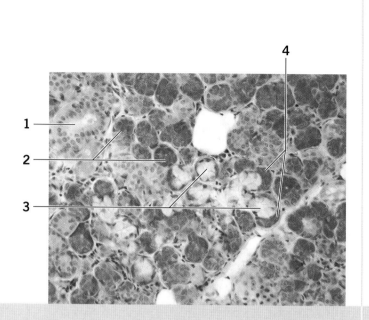

FIGURE 10.21 Submandibular Gland, Cross Section, 400×

1. Duct
2. Serous acini
3. Mucous acini
4. Serous demilunes

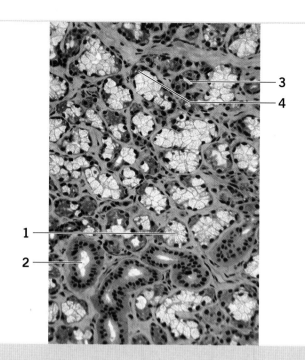

FIGURE 10.22 Sublingual Gland, Cross Section, 400×

1. Mucous acinus
2. Duct
3. Serous acinus
4. Serous demilune

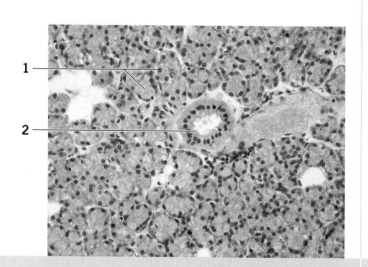

FIGURE 10.23 Parotid Gland, Cross Section, 400×

1. Serous acini
2. Duct

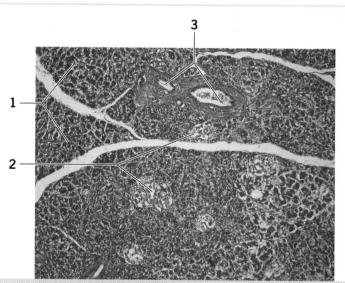

FIGURE 10.24 Pancreas, Cross Section, 100×

1. Acinar tissue
2. Pancreatic islets
3. Ducts

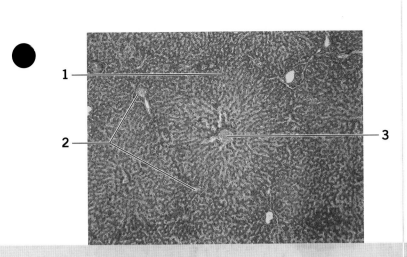

FIGURE 10.25 Liver, Cross Section, 100×

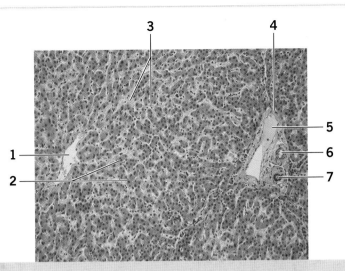

FIGURE 10.26 Liver, Cross Section, 200×

1. Liver lobule
2. Portal triads
3. Central vein

1. Central vein
2. Hepatocytes
3. Sinusoids
4. Portal triad
5. Portal venule
6. Portal arteriole
7. Bile duct

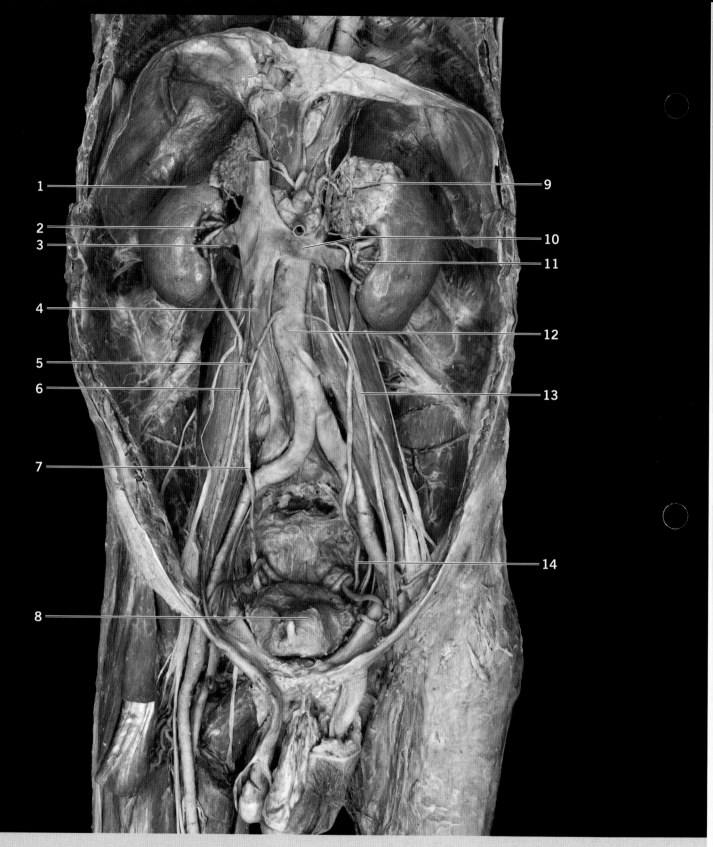

FIGURE 11.1 Structures of the Urinary System

1. Right Kidney
2. Right renal artery
3. Right renal vein
4. Inferior vena cava
5. Right gonadal artery
6. Right gonadal vein
7. Right ureter
8. Urinary bladder
9. Left adrenal gland
10. Left renal vein
11. Left renal artery
12. Abdominal aorta
13. Left gonadal vein
14. Left ureter

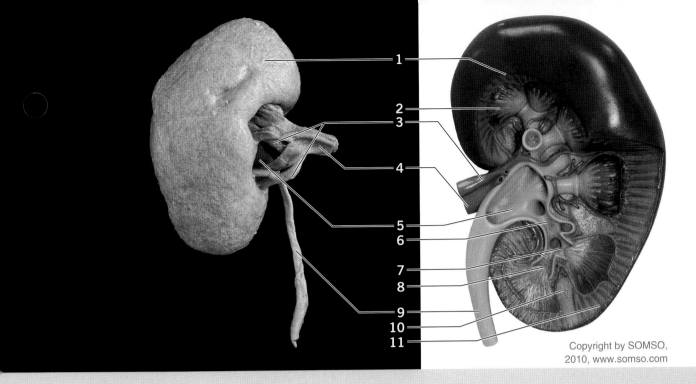

Copyright by SOMSO,
2010, www.somso.com

FIGURE 11.2 Kidney

1. Renal capsule
2. Renal pyramid
3. Renal artery
4. Renal vein
5. Renal pelvis
6. Major calyx
7. Minor calyces
8. Renal papilla
9. Ureter
10. Renal column
11. Renal cortex

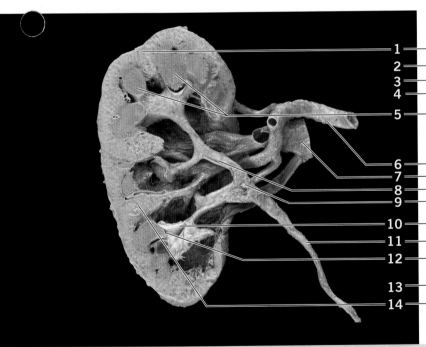

Copyright by SOMSO, 2010, www.somso.com

FIGURE 11.3 Kidney, Frontal Section

1. Renal cortex
2. Arcuate vessels
3. Interlobar vessels
4. Nephron
5. Renal pyramids
6. Renal artery
7. Renal vein
8. Major calyx
9. Renal pelvis
10. Minor calyx
11. Ureter
12. Renal papilla
13. Renal capsule
14. Renal column

Gross Anatomy of the Urinary System

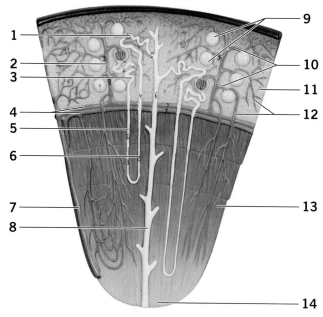

Copyright by SOMSO, 2010, www.somso.com

FIGURE 11.4 Uriniferous Tubule Model

1. Distal convoluted tubule
2. Cortical radiate artery
3. Proximal convoluted tubule
4. Arcuate vessels
5. Descending limb of loop of Henle
6. Ascending limb of loop of Henle
7. Interlobar artery
8. Collecting duct
9. Renal corpuscles
10. Afferent arterioles
11. Renal cortex
12. Efferent arterioles
13. Renal pyramid
14. Renal papilla

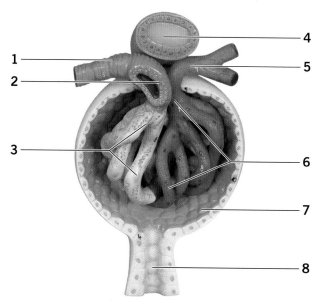

Copyright by SOMSO, 2010, www.somso.com

FIGURE 11.5 Renal Corpuscle Model

1. Afferent arteriole
2. Granular Cell
3. Podocytes
4. Distal convoluted tubule
5. Efferent arteriole
6. Glomerular capillaries
7. Glomerular capsule
8. Proximal convoluted tubule

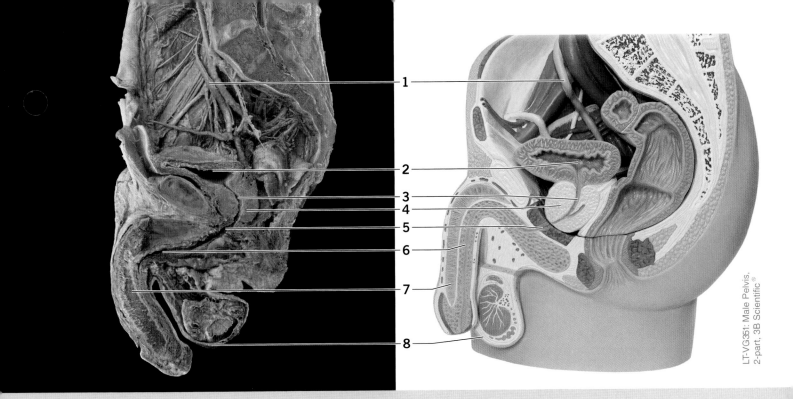

FIGURE 11.6 Urinary Structures of Male Pelvis, Bisected View

1. Ureter
2. Urinary bladder

3. Prostatic urethra
4. *Prostate gland*

5. Membranous urethra
6. Spongy urethra

7. Corpus cavernosum
8. *Scrotum*

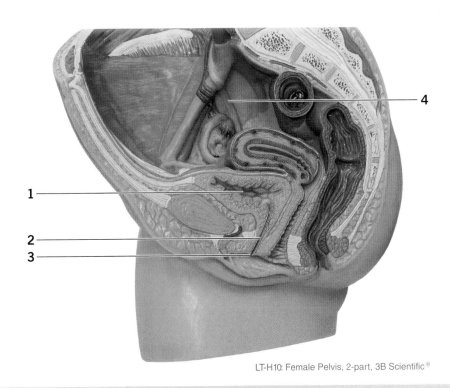

FIGURE 11.7 Urinary Structures of Female Pelvis, Bisected View

1. Urinary bladder

2. Urethra

3. External urethral orifice

4. Ureter

Gross Anatomy of the Urinary System

Chapter 11   The Urinary System   **183**

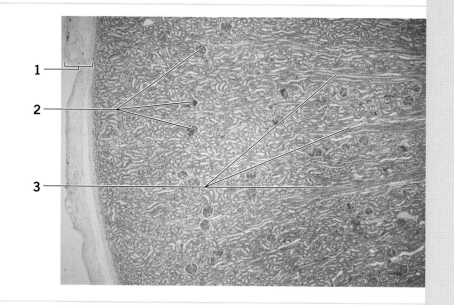

FIGURE 11.8 Renal Cortex, Longitudinal Section, 4×

1. Renal capsule
2. Renal corpuscles
3. Medullary rays

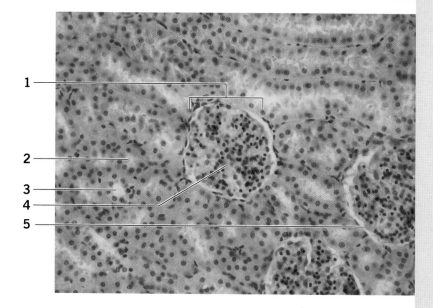

FIGURE 11.9 Renal Cortex, Longitudinal Section, 400×

1. Renal corpuscle
2. Proximal convoluted tubule
3. Distal convoluted tubule
4. Glomerular capillaries
5. Glomerular capsule

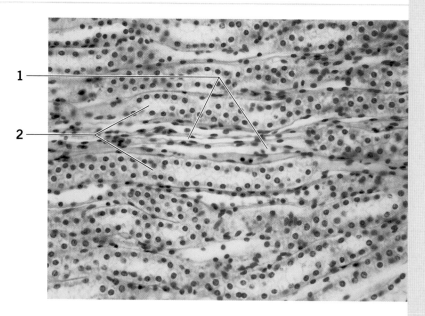

FIGURE 11.10 Renal Medulla, Longitudinal Section, 400×

1. Loops of Henle
2. Collecting ducts

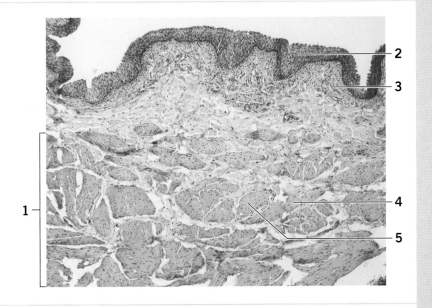

FIGURE 11.11 Urinary Bladder, Cross Section, 100×

1. Muscular layer
2. Transitional epithelium
3. Lamina propria
4. Circular smooth muscle fibers
5. Longitudinal smooth muscle fibers

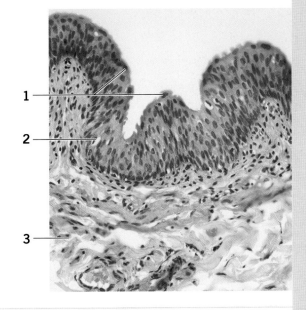

FIGURE 11.12 Urinary Bladder, Cross Section, 400×

1. Umbrella cells
2. Transitional epithelium
3. Lamina propria

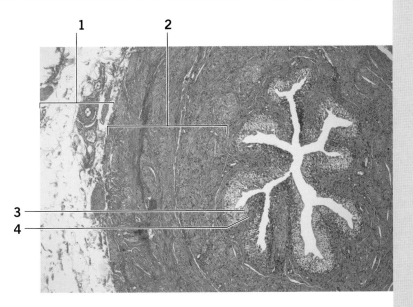

FIGURE 11.13 Ureter, Cross Section, 100×

1. Adventitia
2. Muscular layer
3. Transitional epithelium
4. Lamina propria

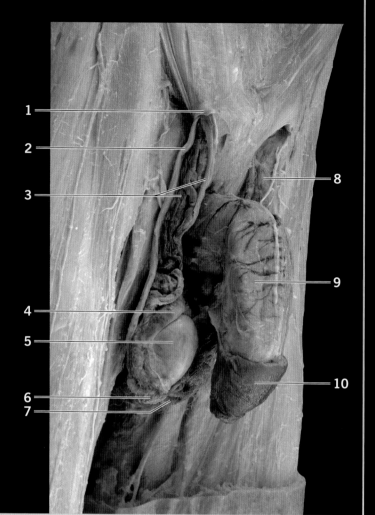

FIGURE 12.1 External Male Genitalia

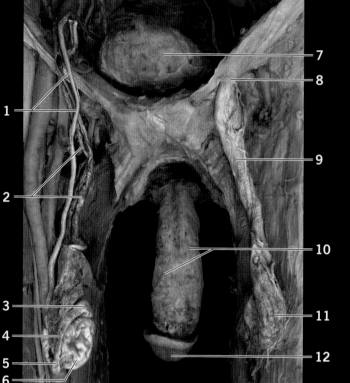

FIGURE 12.2 External Male Genitalia, Dissected

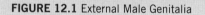

1. Superficial inguinal ring
2. Ductus deferens
3. Testicular artery and pampiniform venous plexus
4. Head of epididymis

5. Testis
6. Tail of epididymis
7. Scrotum
8. Left spermatic cord
9. Penis
10. Prepuce

1. Ductus deferens
2. Testicular artery and pampiniform venous plexus
3. Head of epididymis
4. Body of epididymis
5. Tail of epididymis
6. Testis
7. *Urinary bladder*

8. Superficial inguinal ring
9. Spermatic cord
10. Corpora cavernosa of penis
11. Testis in external spermatic fascia
12. Glans penis

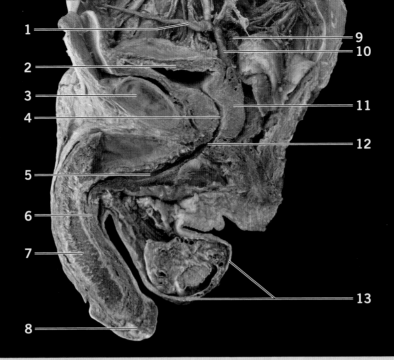

FIGURE 12.3 Male Pelvis, Midsaggital Section

1. Ductus deferens
2. Urinary bladder
3. Pubic symphysis
4. Prostatic urethra
5. Spongy urethra
6. Corpus spongiosum
7. Corpus cavernosum
8. Glans penis
9. Seminal vesicle
10. Ampulla of ductus deferens
11. Prostate gland
12. Membranous urethra
13. Scrotum containing testis and epididymis

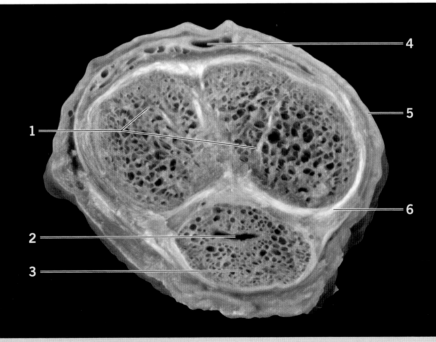

FIGURE 12.4 Penis, Cross Section

1. Corpora cavernosa
2. Spongy urethra
3. Corpus spongiosum
4. Deep dorsal vein
5. Skin
6. Tunica albuginea

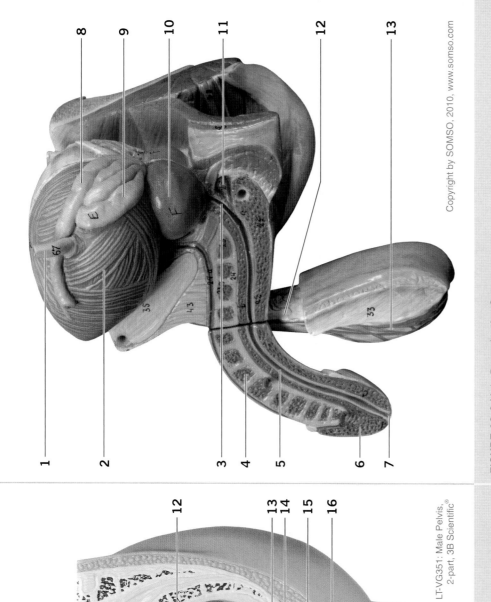

LT-VG351: Male Pelvis,
2-part, 3B Scientific®

FIGURE 12.5 Male Pelvis Model, Midsagittal View

1. *Urinary bladder*
2. *Pubic symphysis*
3. Prostate gland
4. Corpus cavernosum
5. Spongy urethra
6. Corpus spongiosum
7. Prepuce
8. Glans penis
9. External urethral orifice
10. Testis
11. Scrotum
12. Ductus deferens
13. Ejaculatory duct
14. Prostatic urethra
15. Membranous urethra
16. *Anal canal*

Copyright by SOMSO, 2010, www.somso.com

FIGURE 12.6 Male Reproductive Organs

1. *Ureter*
2. *Urinary bladder*
3. Membranous urethra
4. Corpus cavernosum
5. Spongy urethra
6. Glans penis
7. External urethral orifice
8. Ampulla of ductus deferens
9. Seminal vesicle
10. Prostate gland
11. Bulbourethral gland
12. Spermatic cord
13. Testis in scrotum

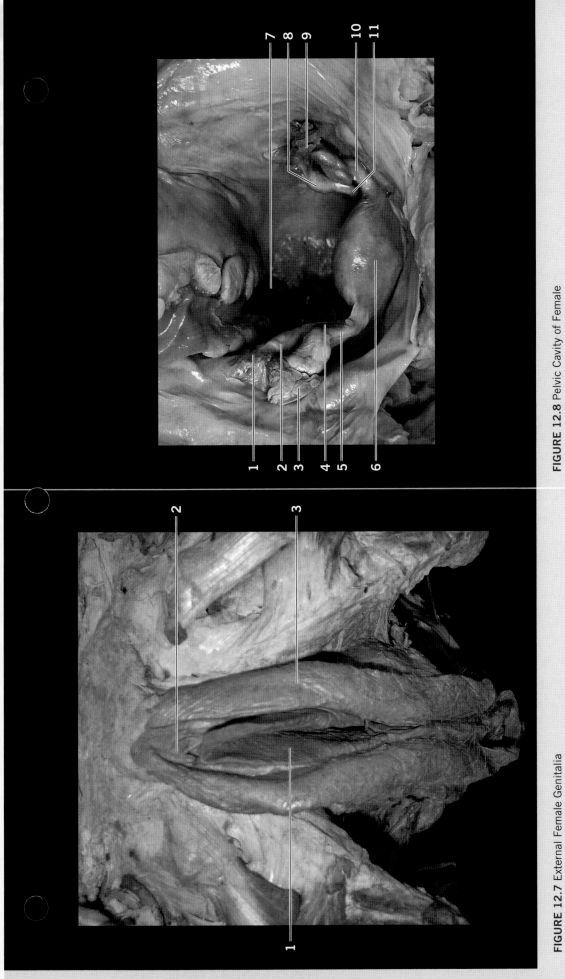

FIGURE 12.8 Pelvic Cavity of Female

1. Suspensory ligament of ovary
2. Right ovary
3. Fimbriae of right uterine tube
4. Right ovarian ligament
5. Right uterine tube
6. Uterus

7. Rectum
8. Left ovary
9. Fimbriae of left uterine tube
10. Left uterine tube
11. Left ovarian ligament

FIGURE 12.7 External Female Genitalia

1. Labium minus
2. Clitoris

3. Labium majus

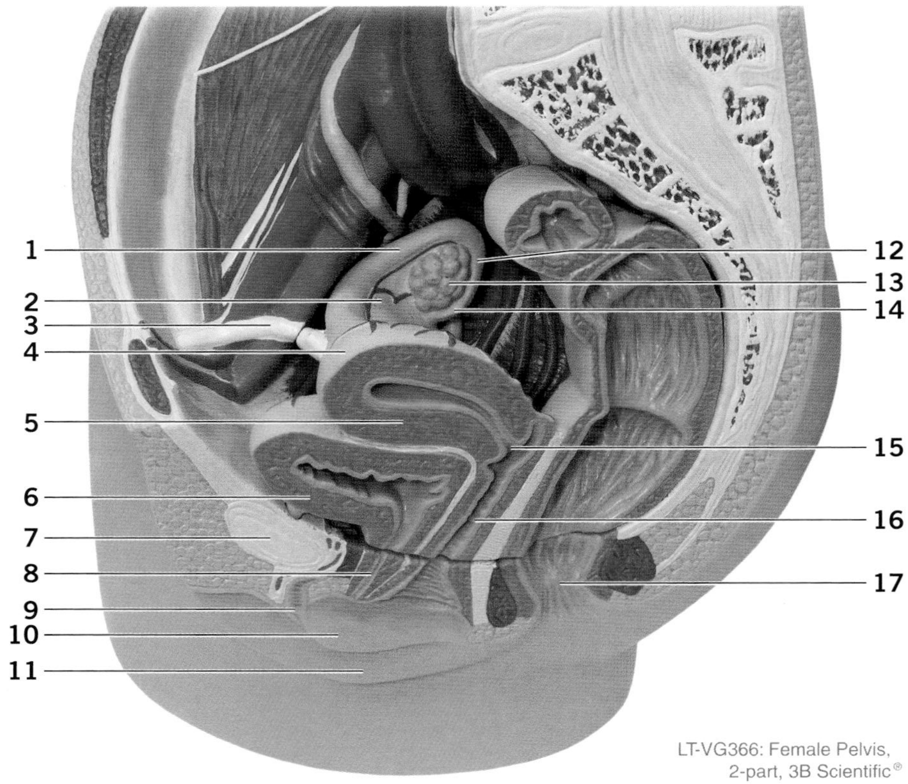

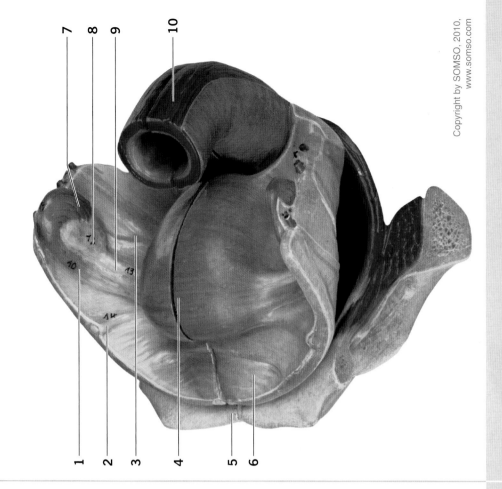

FIGURE 12.9 Female Pelvis Model, Midsagittal View

1. Ampulla of uterine tube
2. Ovarian ligament
3. Round ligament
4. Fundus of uterus
5. Body of uterus
6. *Urinary bladder*
7. *Pubic symphysis*
8. *Urethra*
9. *Clitoris*
10. Labium minus
11. Labium majus
12. Infundibulum of uterine tube
13. Ovary
14. Fimbriae of uterine tube
15. Cervix of uterus
16. Vagina
17. *Anal canal*

FIGURE 12.10 Uterus Model, Laterosuperior View

1. Uterine tube
2. Round ligament
3. Broad ligament
4. Uterus
5. *Pubic symphysis*
6. *Urinary bladder*
7. Fimbriae of uterine tube
8. Ovary
9. Ovarian ligament
10. *Sigmoid colon*

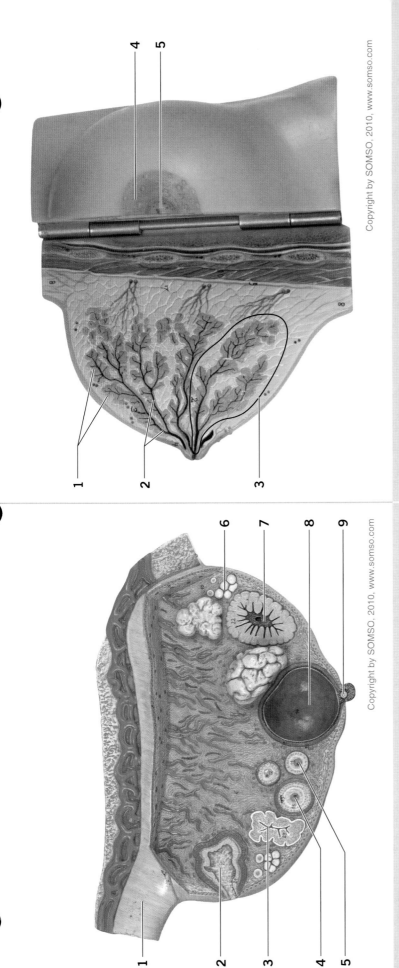

Copyright by SOMSO, 2010, www.somso.com

FIGURE 12.12 Mammary Gland Model

1. Lobules
2. Lactiferous ducts
3. Lobe
4. Areola
5. Nipple

Copyright by SOMSO, 2010, www.somso.com

FIGURE 12.11 Ovary Model, Bisected View

1. Ovarian ligament
2. Ruptured follicle
3. Corpus albicans
4. Secondary follicle
5. Primary follicle
6. Primordial follicle
7. Corpus luteum
8. Mature tertiary (Graafian) follicle
9. Oocyte

Gross Anatomy of the Reproductive System

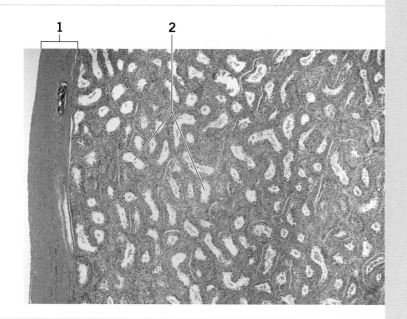

FIGURE 12.13 Testis, Cross Section, 40×

1. Tunica albuginea
2. Seminiferous tubules

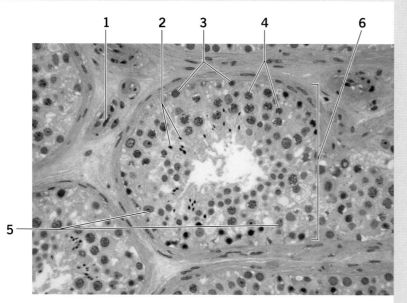

FIGURE 12.14 Seminiferous Tubule of Testis, Cross Section, 400×

1. Interstitial cell
2. Spermatids
3. Spermatogonia
4. Primary spermatocytes
5. Sustentacular cell nuclei
6. Seminiferous tubule

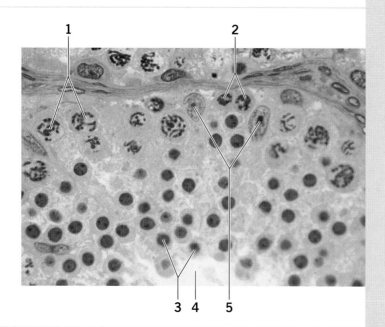

FIGURE 12.15 Seminiferous Tubule of Testis, Cross Section, 1000×

1. Primary spermatocytes
2. Spermatogonia
3. Spermatids
4. Lumen of seminiferous tubule
5. Sustentacular cell nuclei

Tissues of the Reproductive System

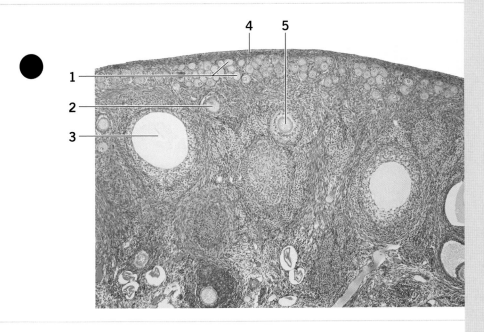

FIGURE 12.16 Ovary, Cross Section, 100×

1. Primordial follicles
2. Primary follicle
3. Antrum
4. Tunica albuginea
5. Secondary follicle

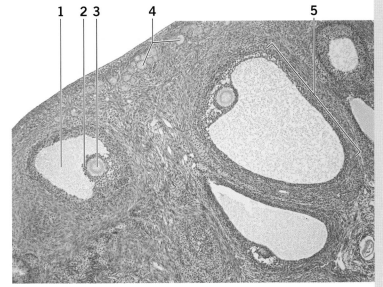

FIGURE 12.17 Structures of a Tertiary Follicle, 100×

1. Antrum
2. Corona radiata
3. Oocyte
4. Primordial follicles
5. Mature tertiary (Graafian) follicle

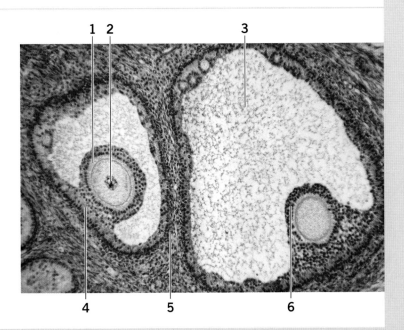

FIGURE 12.18 Mature Tertiary Follicle of Ovary, 200×

1. Zona pellucida
2. Oocyte
3. Antrum
4. Granulosa cells
5. Theca interna
6. Corona radiata

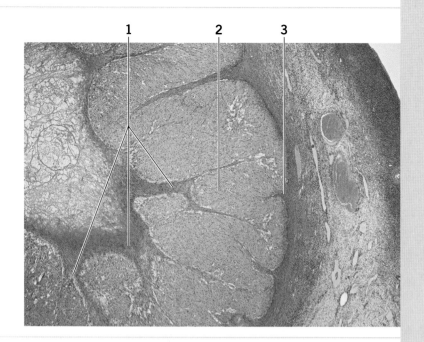

FIGURE 12.19 Corpus Luteum of Ovary, 40×

1. Connective tissue septa
2. Granulosa lutein cells
3. Theca lutein cells

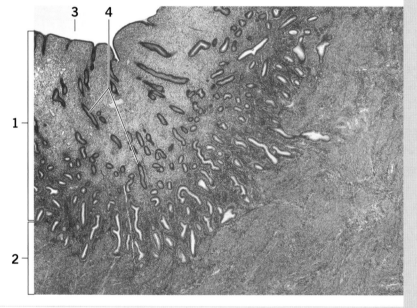

FIGURE 12.20 Uterus During Proliferative Phase, Cross Section, 40×

1. Endometrium
2. Myometrium
3. Lumen of uterus
4. Uterine glands

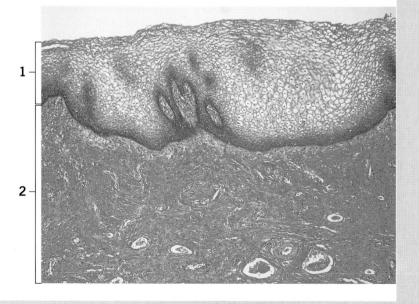

FIGURE 12.21 Vagina, Cross Section, 100×

1. Stratified squamous epithelium
2. Lamina propria

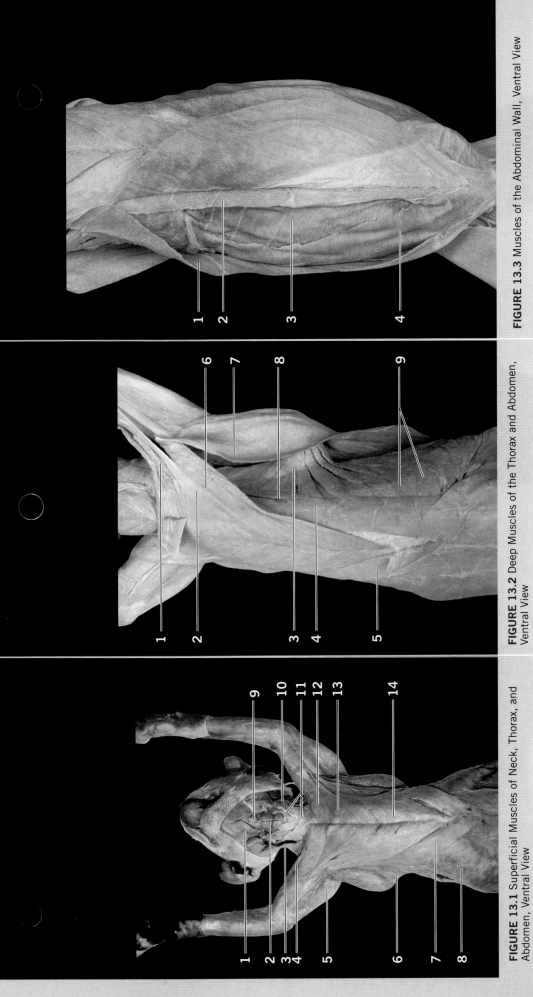

FIGURE 13.1 Superficial Muscles of Neck, Thorax, and Abdomen, Ventral View

1. Digastric
2. Sternohyoid
3. Clavotrapezius
4. Clavobrachialis
5. Triceps brachii
6. Latissimus dorsi
7. Xiphihumeralis
8. External oblique
9. Mylohyoid
10. *External jugular vein*
11. Sternomastoid
12. Pectoantebrachialis
13. Pectoralis major
14. Pectoralis minor

FIGURE 13.2 Deep Muscles of the Thorax and Abdomen, Ventral View

1. Pectoantebrachialis
2. Pectoralis major
3. Serratus ventralis
4. Rectus abdominis
5. Xiphihumeralis
6. Pectoralis minor
7. Latissimus dorsi
8. Scalenus
9. Intercostals

FIGURE 13.3 Muscles of the Abdominal Wall, Ventral View

1. External oblique (cut and reflected)
2. Rectus abdominis
3. Transversus abdominis
4. Internal oblique

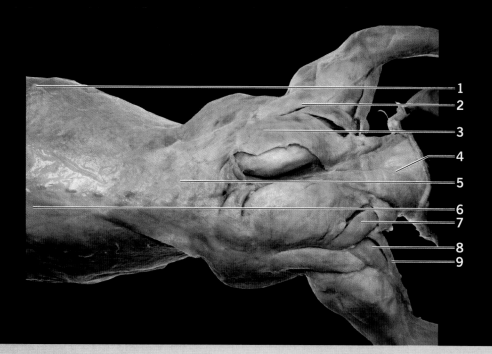

FIGURE 13.4 Superficial Muscles of the Torso, Dorsal View

1. Latissimus dorsi
2. Spinodeltoid
3. Acromiotrapezius
4. Clavotrapezius
5. Spinotrapezius
6. Lumbodorsal fascia
7. Levator scapulae ventralis
8. Acromiodeltoid
9. Clavodeltoid

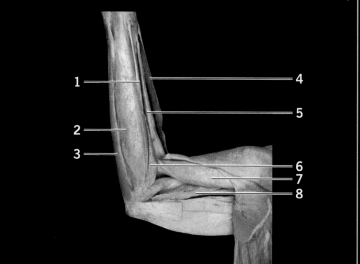

FIGURE 13.5 Muscles of the Forelimb, Medial View, Right Side

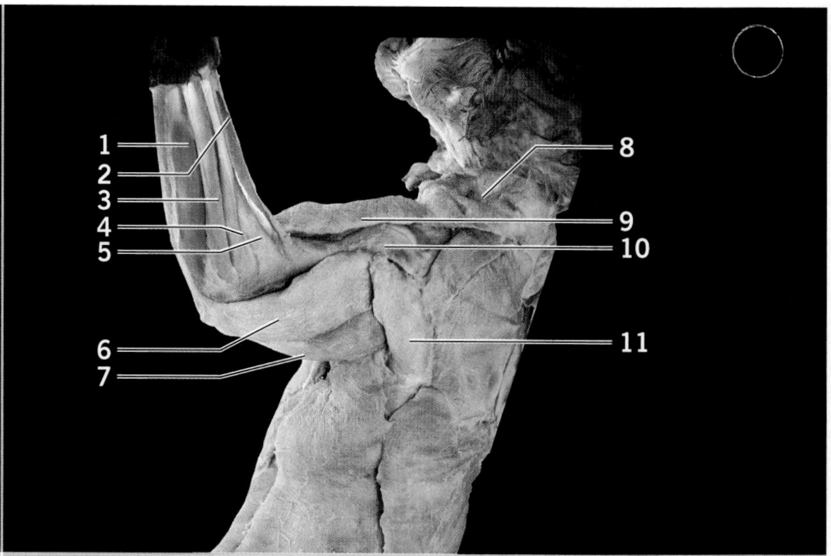

FIGURE 13.6 Muscles of the Forelimb, Lateral View, Left Side

1. Flexor carpi radialis
2. Palmaris longus
3. Flexor carpi ulnaris
4. Brachioradialis
5. Extensor carpi radialis
6. Pronator teres
7. Biceps brachii
8. Triceps brachii (medial head)

1. Extensor carpi ulnaris
2. Brachioradialis
3. Extensor digitorum lateralis
4. Extensor digitorum communis
5. Extensor carpi radialis longus
6. Triceps brachii (lateral head)
7. Triceps brachii (long head)
8. Clavotrapezius
9. Clavodeltoid
10. Acromiodeltoid
11. Spinodeltoid

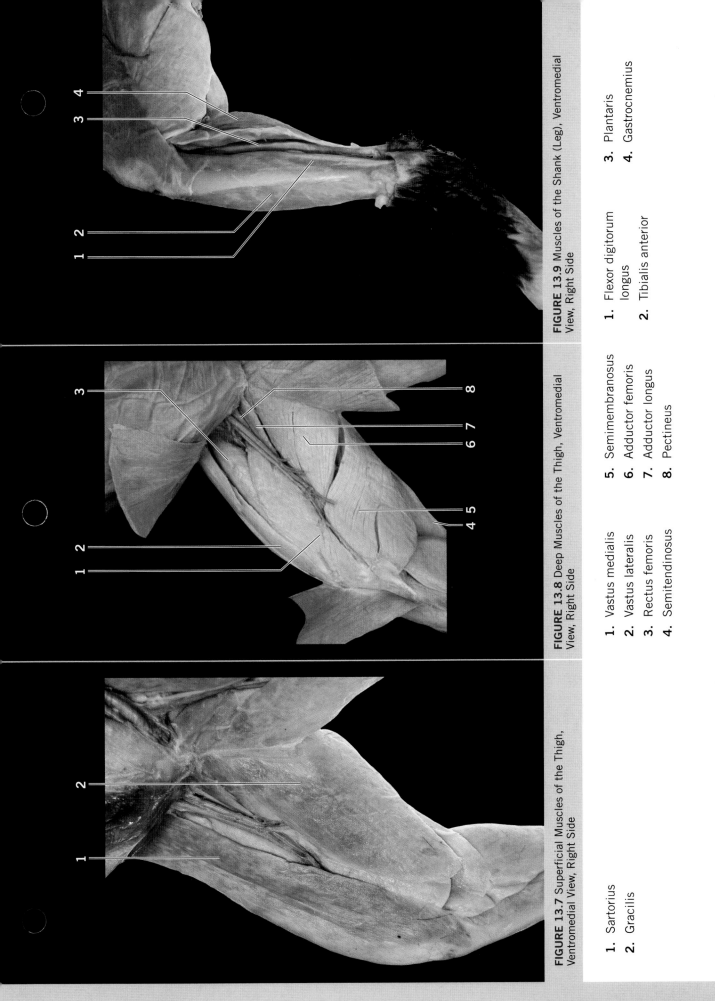

FIGURE 13.7 Superficial Muscles of the Thigh, Ventromedial View, Right Side

1. Sartorius
2. Gracilis

FIGURE 13.8 Deep Muscles of the Thigh, Ventromedial View, Right Side

1. Vastus medialis
2. Vastus lateralis
3. Rectus femoris
4. Semitendinosus
5. Semimembranosus
6. Adductor femoris
7. Adductor longus
8. Pectineus

FIGURE 13.9 Muscles of the Shank (Leg), Ventromedial View, Right Side

1. Flexor digitorum longus
2. Tibialis anterior
3. Plantaris
4. Gastrocnemius

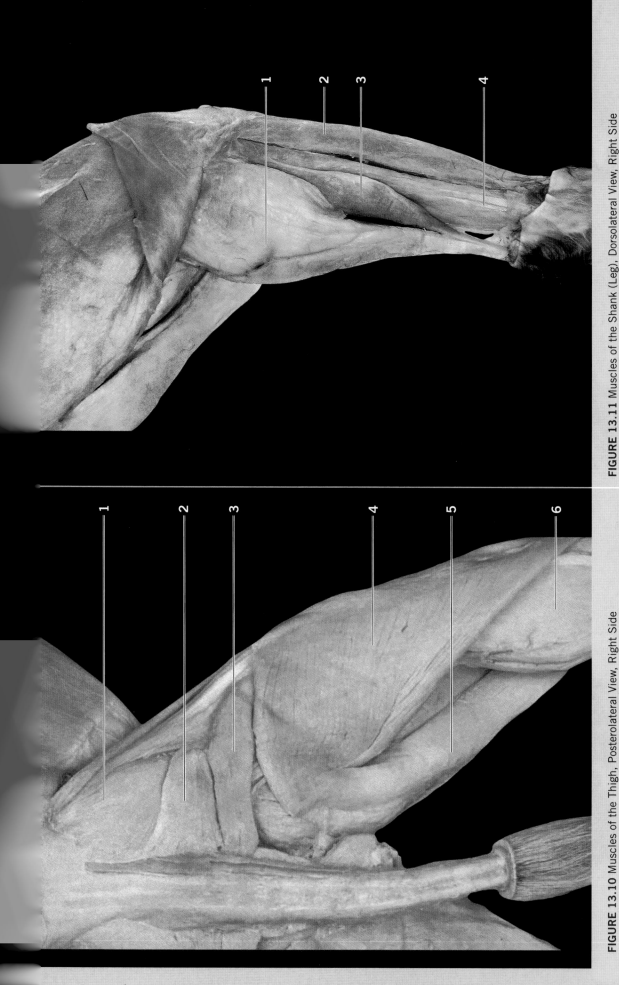

FIGURE 13.11 Muscles of the Shank (Leg), Dorsolateral View, Right Side

1. Gastrocnemius
2. Extensor digitorum longus
3. Soleus
4. Fibularis muscles

FIGURE 13.10 Muscles of the Thigh, Posterolateral View, Right Side

1. Gluteus medius
2. Gluteus maximus
3. Caudofemoralis
4. Biceps femoris
5. Semitendinosus
6. Gastrocnemius

Muscular System

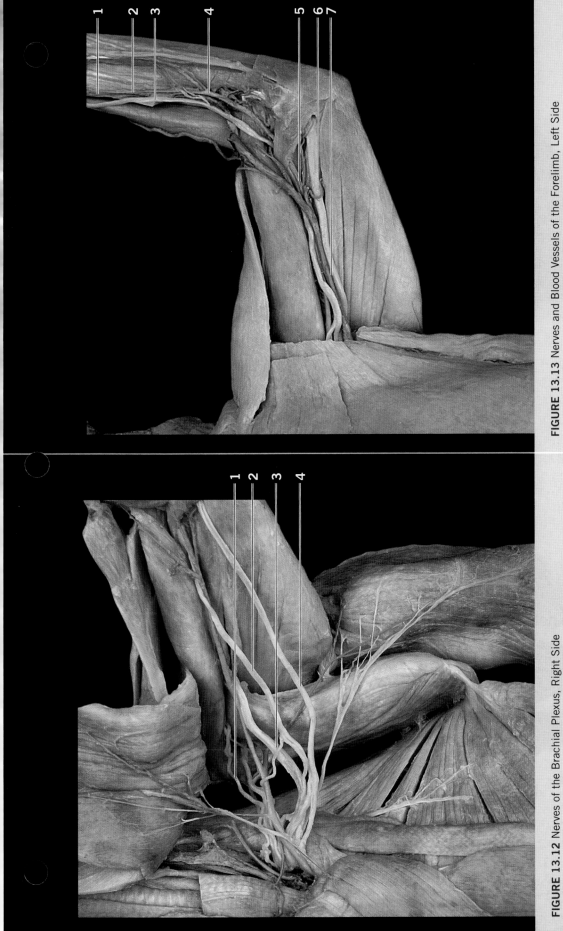

FIGURE 13.13 Nerves and Blood Vessels of the Forelimb, Left Side

1. Radial vein
2. Radial artery
3. Median nerve
4. Ulnar artery

5. Brachial artery
6. Ulnar nerve
7. Brachial vein

FIGURE 13.12 Nerves of the Brachial Plexus, Right Side

1. Musculocutaneous nerve
2. Median nerve

3. Radial nerve
4. Ulnar nerve

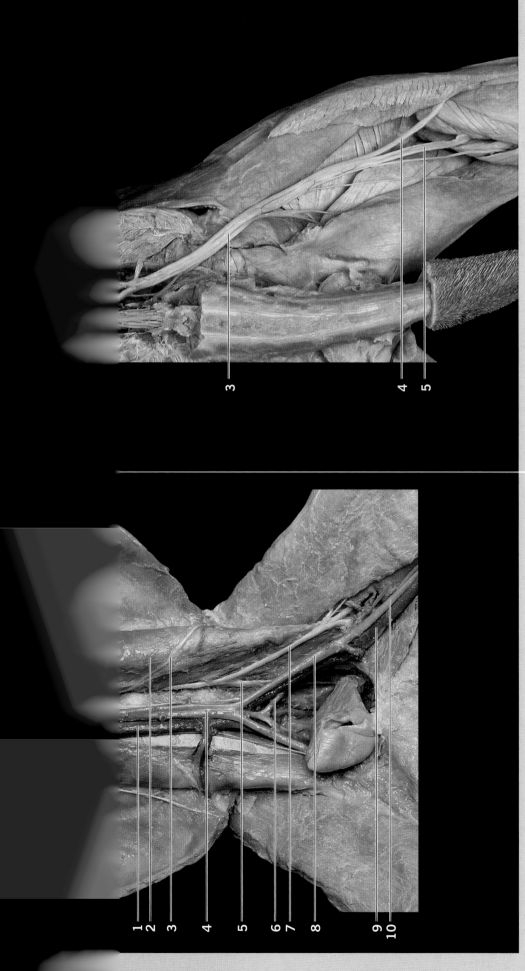

FIGURE 13.14 Nerves of the Lumbar Plexus and Blood Vessels of the Pelvis, Left Side

1. Inferior vena cava
2. *Iliopsoas muscle*
3. Lateral femoral cutaneous nerve
4. Descending aorta
5. Obturator nerve

6. Internal iliac artery
7. Femoral nerve
8. External iliac artery
9. Femoral vein
10. Femoral artery

FIGURE 13.15 Nerves of the Sacral Plexus

1. *Spinal cord*
2. *Cauda equina*
3. Sciatic nerve

4. Common fibular nerve
5. Tibial nerve

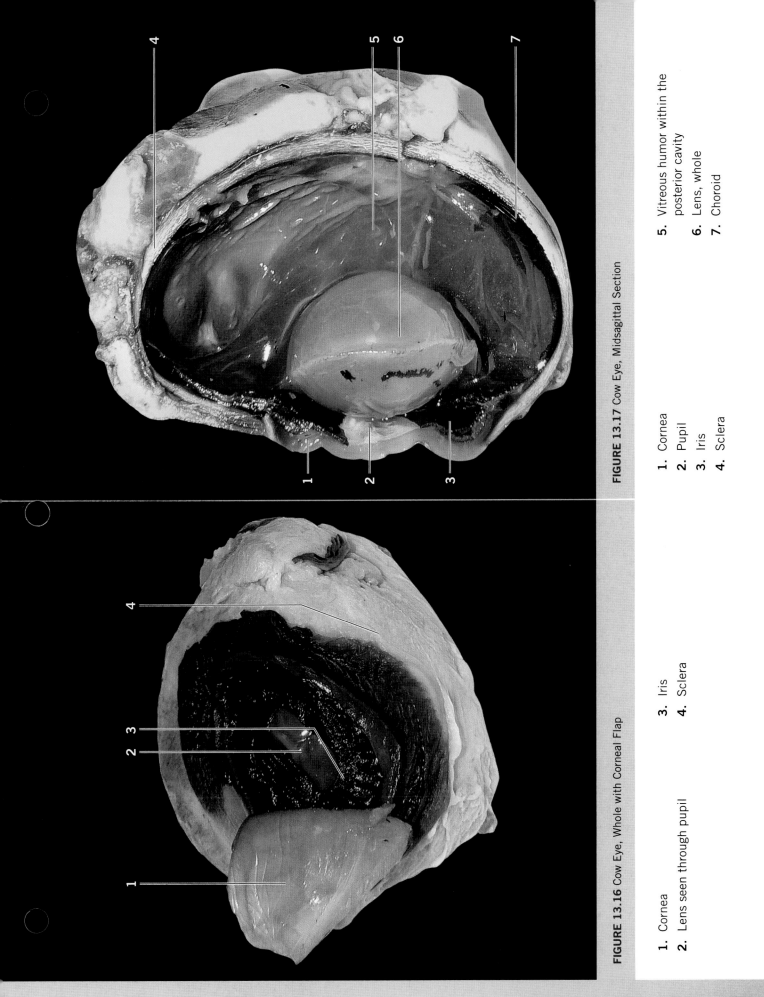

FIGURE 13.17 Cow Eye, Midsagittal Section

1. Cornea
2. Pupil
3. Iris
4. Sclera

5. Vitreous humor within the posterior cavity
6. Lens, whole
7. Choroid

FIGURE 13.16 Cow Eye, Whole with Corneal Flap

1. Cornea
2. Lens seen through pupil

3. Iris
4. Sclera

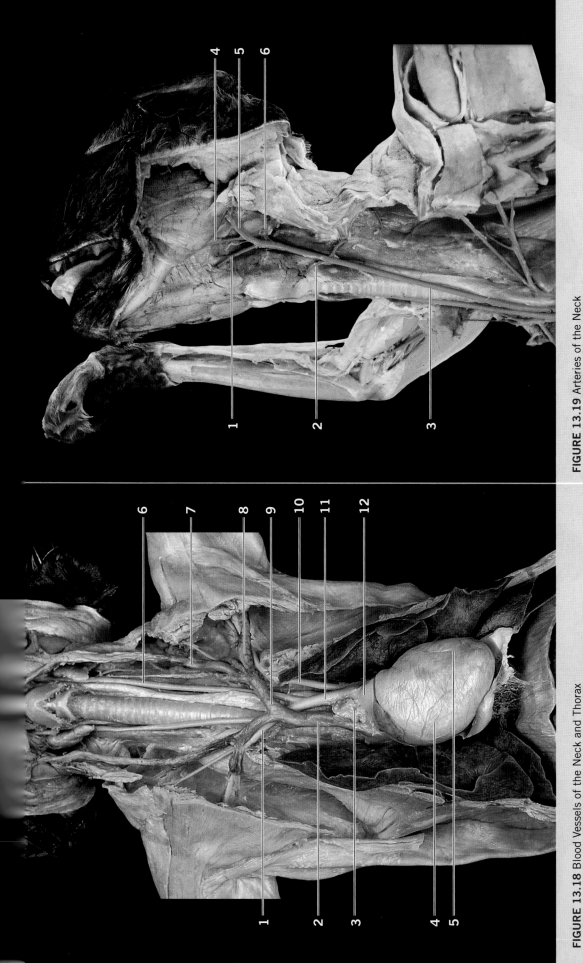

FIGURE 13.18 Blood Vessels of the Neck and Thorax

1. Right subclavian artery
2. Superior vena cava
3. *Thymus*
4. *Right ventricle*
5. *Left ventricle*
6. Left common carotid artery
7. Left external jugular vein
8. Left subclavian vein
9. Left brachiocephalic vein
10. Subclavian artery
11. Brachiocephalic trunk
12. Aortic arch

FIGURE 13.19 Arteries of the Neck

1. Lingual artery
2. Superior thyroid artery
3. Common carotid artery
4. External maxillary artery
5. External carotid artery
6. Internal carotid artery

Cardiovascular System

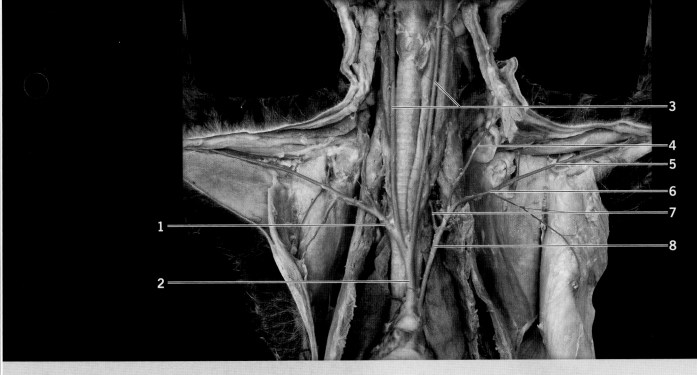

FIGURE 13.20 Arteries of the Neck and Thorax

1. Right subclavian artery
2. Brachiocephalic trunk
3. Common carotid arteries
4. Left thyrocervical artery
5. Left brachial artery
6. Left axillary artery
7. Left vertebral artery
8. Left subclavian artery

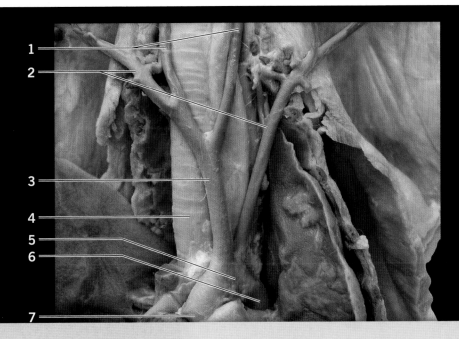

FIGURE 13.21 Branches of the Aortic Arch

1. Common carotid arteries
2. Subclavian arteries
3. Brachiocephalic trunk
4. *Trachea*
5. Aortic arch
6. Descending aorta
7. Ascending aorta

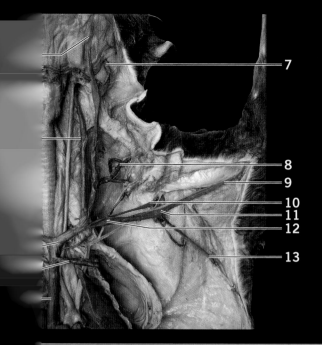

FIGURE 13.22 Veins of the Face, Neck, and Thorax

1. Anterior facial vein
2. Transverse jugular vein
3. Internal jugular veins
4. Brachio-cephalic veins
5. Internal mammary veins
6. Superior vena cava

7. Posterior facial vein
8. Transverse scapular vein
9. Brachial vein
10. Subscapular vein
11. Axillary vein
12. Subclavian vein
13. Long thoracic vein

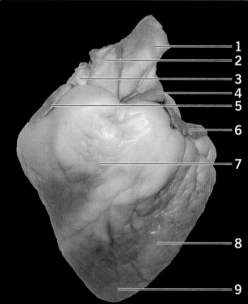

FIGURE 13.23 Sheep Heart, Ventral View

1. Aorta
2. Brachio-cephalic trunk
3. Superior vena cava
4. Pulmonary trunk

5. Right atrium
6. Left atrium
7. Right ventricle
8. Left ventricle
9. Apex of the heart

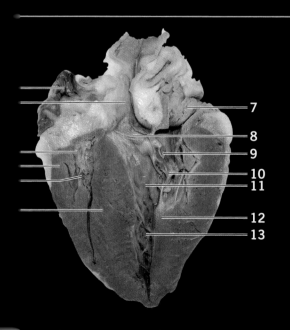

FIGURE 13.24 Sheep Heart, Frontal Section

1. Right atrium
2. Aorta
3. Tricuspid valve
4. Myocardium
5. Right ventricle
6. Interventri-cular septum
7. Left atrium

8. Aortic semilunar valve
9. Bicuspid valve
10. Chordae tendineae
11. Left ventricle
12. Papillary muscle
13. Trabeculae carneae

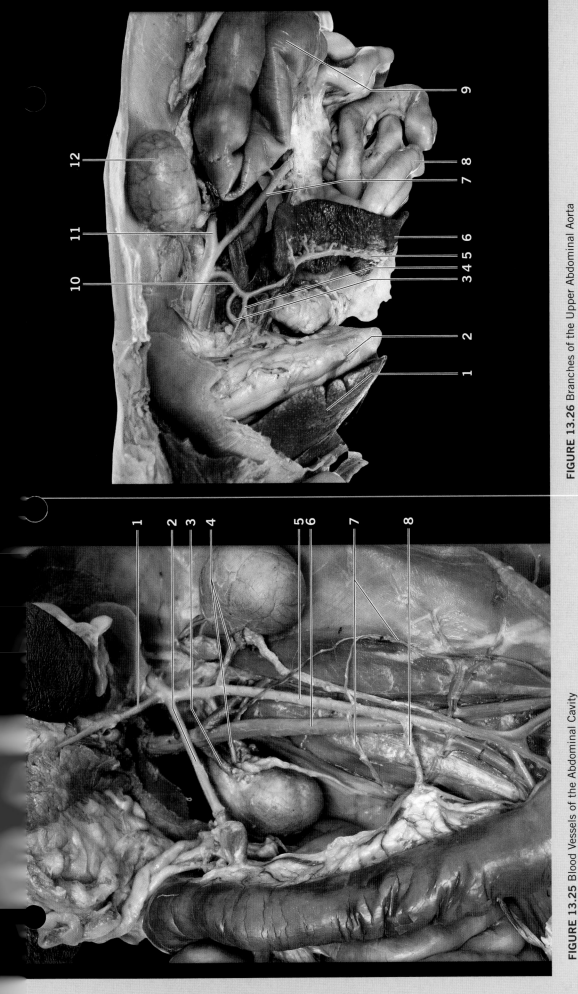

FIGURE 13.25 Blood Vessels of the Abdominal Cavity

1. Celiac trunk
2. Superior mesenteric artery
3. Right renal vein
4. Renal arteries
5. Descending aorta
6. Inferior vena cava
7. Right gonadal artery
8. Inferior mesenteric artery

FIGURE 13.26 Branches of the Upper Abdominal Aorta

1. *Liver*
2. *Stomach*
3. Hepatic artery
4. Left gastric artery
5. Splenic artery
6. *Spleen*
7. Superior mesenteric artery
8. *Small intestine*
9. *Large intestine*
10. Celiac trunk
11. Descending aorta
12. *Left kidney*

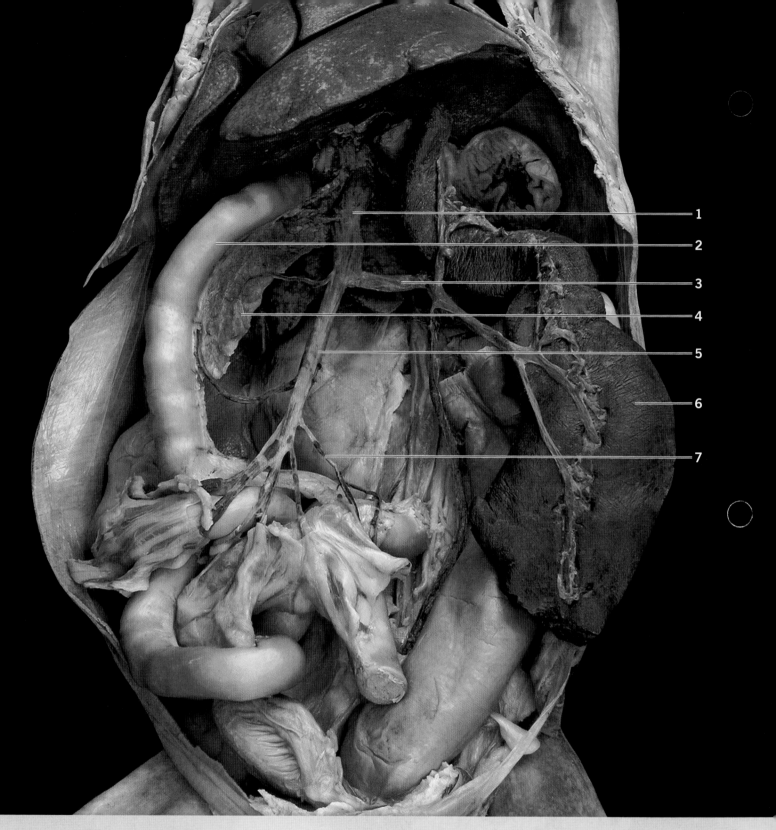

FIGURE 13.27 Hepatic Portal System

1. Hepatic portal vein
2. Duodenum
3. Gastrosplenic vein
4. Pancreas
5. Superior mesenteric vein
6. Spleen
7. Inferior mesenteric vein

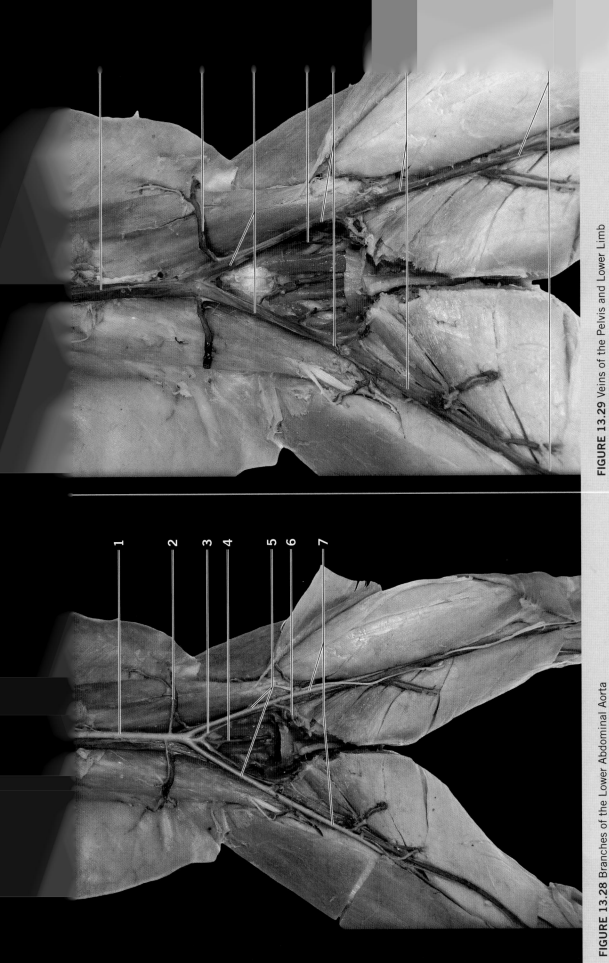

FIGURE 13.28 Branches of the Lower Abdominal Aorta

1. Descending aorta
2. Left iliolumbar artery
3. Left internal iliac artery
4. Median sacral artery
5. External iliac arteries
6. Left deep femoral artery
7. Femoral arteries

FIGURE 13.29 Veins of the Pelvis and Lower Limb

1. Inferior vena cava
2. Left iliolumbar vein
3. Common iliac veins
4. Left internal iliac vein
5. External iliac veins
6. Femoral veins
7. Great saphenous veins

Cardiovascular System

Chapter 13 Cat Dissections

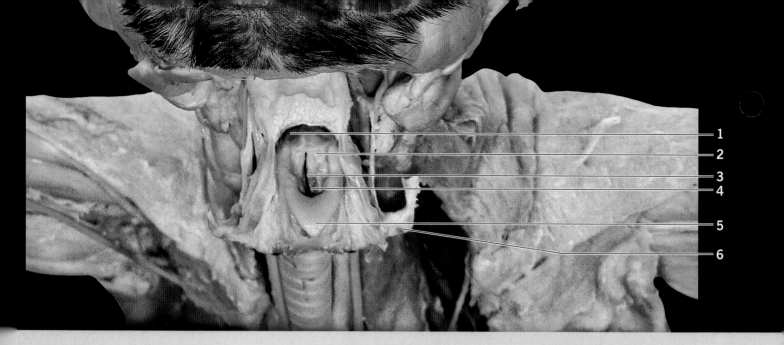

FIGURE 13.30 Larynx, Superior View

1. Opening to esophagus
2. Vestibular fold
3. Glottis
4. True vocal fold (cord)
5. Epiglottis
6. Hyoid bone

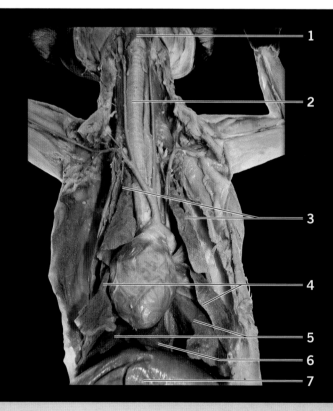

FIGURE 13.31 Respiratory Organs, Ventral View

1. Larynx
2. Trachea
3. Anterior lobes of lung
4. Middle lobes of lung
5. Posterior lobes of lung
6. Mediastinal lobe of lung, right side
7. Respiratory diaphragm

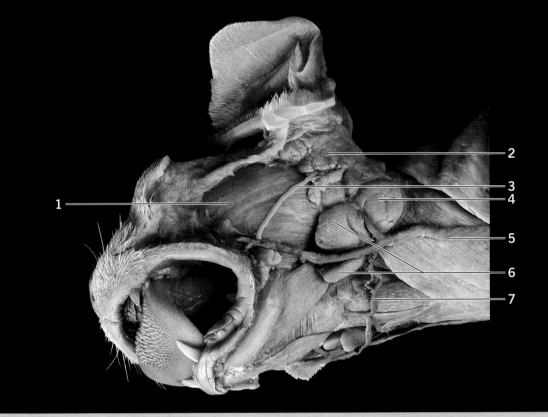

FIGURE 13.32 Salivary Glands

1. *Masseter muscle*
2. Parotid gland
3. Parotid duct
4. Submandibular gland
5. *Left external jugular vein*
6. *Lymph nodes*
7. *Transverse jugular vein*

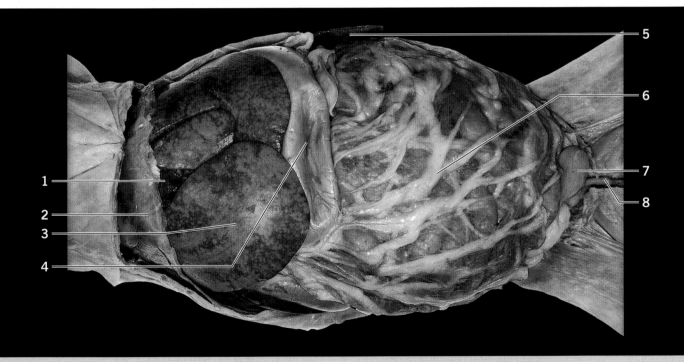

FIGURE 13.33 Abdominal Cavity, Intact

1. Gallbladder
2. Respiratory diaphragm
3. Liver
4. Stomach
5. Spleen
6. Greater omentum
7. Urinary bladder
8. Urethra

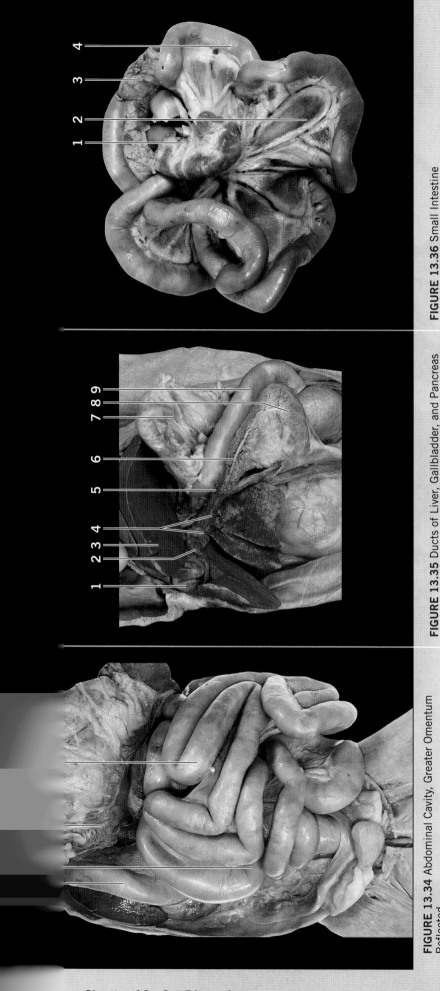

FIGURE 13.34 Abdominal Cavity, Greater Omentum Reflected

1. Duodenum
2. Cecum
3. Pancreas
4. Small intestine

FIGURE 13.35 Ducts of Liver, Gallbladder, and Pancreas

1. Gallbladder
2. Cystic duct
3. Liver
4. Hepatic ducts
5. Bile duct

6. Main pancreatic duct
7. Stomach
8. Pancreas
9. Duodenum

FIGURE 13.36 Small Intestine

1. Lymph node
2. Mesentery
3. Pancreas
4. Small intestine

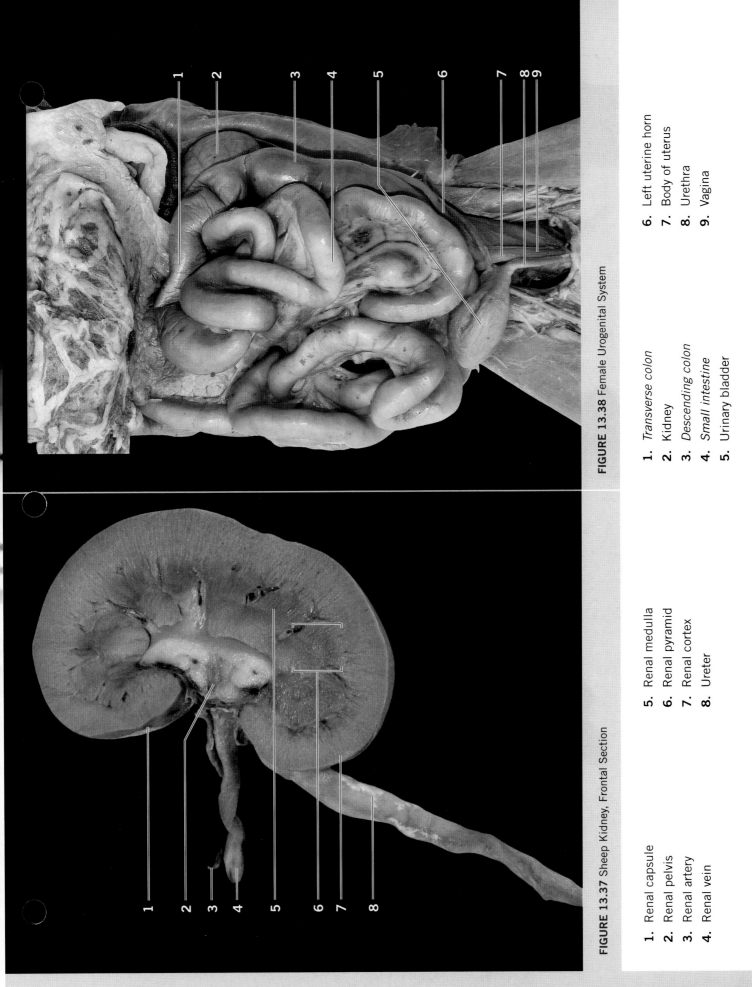

FIGURE 13.38 Female Urogenital System

1. *Transverse colon*
2. Kidney
3. *Descending colon*
4. *Small intestine*
5. Urinary bladder
6. Left uterine horn
7. Body of uterus
8. Urethra
9. Vagina

FIGURE 13.37 Sheep Kidney, Frontal Section

1. Renal capsule
2. Renal pelvis
3. Renal artery
4. Renal vein
5. Renal medulla
6. Renal pyramid
7. Renal cortex
8. Ureter

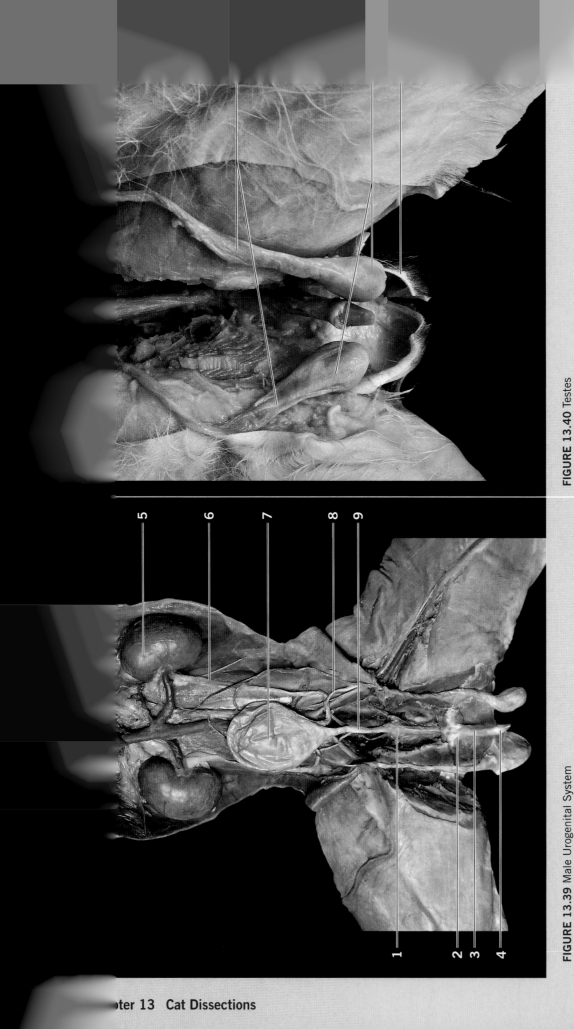

FIGURE 13.40 Testes

1. Spermatic cords
2. Testes
3. Scrotum

FIGURE 13.39 Male Urogenital System

1. Prostate gland
2. Crus of penis
3. Body of penis
4. Glans penis
5. Kidney
6. Ureter
7. Urinary bladder
8. Ductus deferens
9. Urethra

Reproductive System

INDEX

CREDITS

All photos were taken for Pearson PAL™ 3.0 by the following photographers:

Bob Bucella
Samuel Chen

Larry DeLay
Elena Dorfman
Steve Downing
Marjorie Johnson
Karen Krabbenhoft
Lisa Lee

Olga Malakhova
Shawn Miller
Mark Nielsen
Eksel Perez
Winston Charles Poulton

Leif Saul
Byron Spencer
Yvonne Baptiste Syzmanski
Peter Westra
Nina Zanetti